# Do Plants Know Math?

# Do Plants Know Math?

## Unwinding the Story of Plant Spirals, from Leonardo da Vinci to Now

STÉPHANE DOUADY

JACQUES DUMAIS

CHRISTOPHE GOLÉ

NANCY PICK

PRINCETON UNIVERSITY PRESS  ◉  PRINCETON AND OXFORD

Copyright © 2024 by Princeton University Press

Princeton University Press is committed to the protection of copyright and the intellectual property our authors entrust to us. Copyright promotes the progress and integrity of knowledge. Thank you for supporting free speech and the global exchange of ideas by purchasing an authorized edition of this book. If you wish to reproduce or distribute any part of it in any form, please obtain permission.

Requests for permission to reproduce material from this work should be sent to permissions@press.princeton.edu

Published by Princeton University Press
41 William Street, Princeton, New Jersey 08540
99 Banbury Road, Oxford OX2 6JX

press.princeton.edu

All Rights Reserved

ISBN 9780691158655
ISBN (e-book) 9780691261089

British Library Cataloging-in-Publication Data is available

Editorial: Diana Gillooly
Production Editorial: Jaden Young
Text and Cover Design: Chris Ferrante
Production: Erin Suydam
Publicity: Matthew Taylor and Kathryn Stevens
Copyeditor: Laurel Anderton

Cover image: Photograph of a fern (*Polypodiopsida*) by Victor Mozqueda

This book has been composed in Monotype Century Schoolbook and The Future

Printed on acid-free paper. ∞

Printed in China

10 9 8 7 6 5 4 3 2 1

Phyllotaxis is so appealing. It seems accessible, like pi—so many people have tried to "explain" pi. And it's the same with phyllotaxis. Everybody goes out and says, "I'm going to solve the riddle." And every 10 years someone puts out a theory, and often it's a step backward.

**—Jacques Dumais**

One thing I've learned in the woods is that there is no such thing as random. Everything is steeped in meaning, colored by relationships, one thing with another.

**—Robin Wall Kimmerer, *Braiding Sweetgrass***

**SD:** For all the eyes waiting to see, all the brains waiting to marvel
**JD:** To my daughter, may you also rejoice in the beauty that surrounds us
**CG:** To my wife, Liz, and daughter, Marguerite, for their loving support
**NP:** For my beautiful mom, the designer

# CONTENTS

*Preface*   xi

**INTRODUCTION**   1

**PART I. Who Noticed First?**   19
**CHAPTER 1.** The (So-Called) Fibonacci Sequence in History   21
**CHAPTER 2.** Plant Patterns in Leonardo's Notebook   34
**CHAPTER 3.** The Golden Ratio as a New Year's Gift   41

**PART II. Could Early Scientists Explain Plant Spirals?**   51
**CHAPTER 4.** First Spirals in the Dew   53
**CHAPTER 5.** Biomathematics on a Watch Face   68
**CHAPTER 6.** So Many Spirals on a Pinecone   78
**CHAPTER 7.** Irrational Angles in a French Garden   96

**PART III. What Did the Microscope Reveal?**   117
**CHAPTER 8.** A Glimpse of the Growing Tip   119
**CHAPTER 9.** Biomechanics under the Lens   129
**CHAPTER 10.** A Critical Tree on Graph Paper   142

**PART IV. Have Computers Shed Any Light?**     159
**CHAPTER 11.** Sunflowers on Turing's Primitive Computer     161
**CHAPTER 12.** Leaves and Petals as Data Points     175
**CHAPTER 13.** The Big Experiment with Tiny Droplets     183
**CHAPTER 14.** Zigzag Fronts in an Artichoke     201
**CHAPTER 15.** Self-Repeating Patterns in Plants (Perhaps?)     219
**CHAPTER 16.** Leaf Bud *Kirigami*     233

**PART V. What Do Biologists Think?**     245
**CHAPTER 17.** The Hormone That Makes Spirals     247
**CHAPTER 18.** A Cell Division Discovery via Soap Bubbles     254
**CHAPTER 19.** A Brief Detour to Animals     265

**PART VI. Conclusion**     275
**CHAPTER 20.** *Do* Plants Know Math?     277
**CHAPTER 21.** A Spiral Dinner (with Recipes)     281

*Appendix*     289
*Notes*     309
*Illustration Credits*     327
*Acknowledgents*     331
*Index*     333

# PREFACE

**Stéphane Douady**

I was inoculated with patterns when I was a child—spying on the pictures of structures from my architect father and the electron microscope images from my plant biologist mother. And I have never stopped looking around at all Nature offers, ever since my mother asked me when I was five to collect samples for her natural science classes (once I came back holding a live wasp in my hand). I wanted (and still want) to keep learning, so I always knew I wanted to do research, but not biology because my mother's experiments seemed much too hard. Mathematics was not an option either, to avoid the shadow of my famous uncle. My brother-in-law suggested I should do physics, since I knew how to fix a toaster. I took one look at a physics lab and fell in love with experiments on soap bubbles and their interference colors. I have never regretted my decision, especially since I eventually returned to studying (this time through the lens of physics) the lives of plants.

**Jacques Dumais**

On a recent trip to my hometown, Montréal, I stumbled across long-forgotten drawings of monsters and mythical beasts. I had drawn them while in grade school, using only circles, squares, and other geometric shapes. A bloodthirsty caterpillar made of regularly spaced circles is the earliest tangible evidence that I would someday be interested in phyllotaxis and other mathematical features of living organisms. More than a decade separates my childhood artwork from my first serious encounter with the problems posed by plant development. I still remember my excitement when I discovered the possibility of reproducing the growth of a sunflower head from a few simple geometric rules. Not surprisingly, my first scientific work pertaining to phyllotaxis focused on the mechanical forces present in sunflowers. Although my own growth as a scientist led me to more fine-grained questions about plant development, I still feel the same sense of excitement whenever I come across new instances of developmental processes whose essence can be captured with simple mathematical models.

**Christophe Golé**
While in high school in Algeria, I got in trouble with my math teacher. During school breaks, I'd always take off and hitchhike in the Sahara with friends and only make my way back slowly, my eyes full of stars and sand, and my mind anywhere but on equations. In 11th grade, my math teacher told me that I would be held back, refusing to let me move on to the next level. I begged him to give me a chance with an exam. For two weeks, I crammed harder than I ever had and . . . suddenly I got what math was all about! Luckily I passed. My French dad's job had taken us to Algeria, and I wanted to become an agronomist to solve the world hunger problem. But by mistake after high school I ended up in a Paris program geared toward engineering. Although I still could not do calculus properly, I was introduced to topology and loved it. I ended up doing a math PhD in the United States, my mom's home country. Later, in a graduate course I taught on dynamical systems, my student Scott Hotton proposed phyllotaxis and the Mandelbrot set as his final project. The topic tied together plants, dynamical systems, geometry, and topology. I was hooked, and I have been ever since.

**Nancy Pick**
I confess that we named our younger child Milo, after the kid in *The Phantom Tollbooth* who sets off on an adventure into words and numbers—bumping into such characters as the Whether Man and the Mathemagician. Driven by that

same sense of curiosity, I started out as a newspaper reporter and wrote two books about natural history (dinosaur tracks, anyone?). Then one evening 14 years ago, I met Chris the mathematician at a cocktail party thrown by a mutual friend, conceptual artist Tom Friedman (whose sculptures often had a mathematical twist). I told Chris I was looking for a good botanical story for my next book. The moment he described his research on Fibonacci spirals in plants, my eyes lit up. Admittedly, when he convinced his friends Stéphane and Jacques to join us, the project took on a daunting complexity. But just as Milo hoped as he sped off from the tollbooth in his little car, it has turned out to be a very "interesting game," in every way imaginable.

*Carbon offsets have been purchased for this book from Tradewater, which collects and destroys old chlorofluorocarbon stockpiles.*

# INTRODUCTION

Phyllotaxis is an adventure in curiosity. It will lead you on a journey into the wild world where nature meets numbers. You will discover in the pages of this book (if you don't already know) that your life is surrounded by botanical sequences and spirals, and that they are very beautiful.

Along the way, you may find that

- You start seeing patterns in such everyday items as strawberries, pineapples, corn, red cabbage, and artichokes.
- When walking in the woods, you become mesmerized by pinecones.
- You catch yourself counting 1, 1, 2, 3, 5, 8, 13, 21, . . . on a regular basis and finding in that a kind of thrill. Amazingly, the Fibonacci sequence—obtained by adding together the last two numbers to obtain the next one—appears in a vast number of plants.

This will be no dry scientific treatise, but instead a very human adventure. In this book, you will delve into the hearts and minds of scientists who have unlocked the secrets of "plant-mathematics" over the course of several centuries. Leonardo da Vinci makes an appearance, and so does Alan Turing. Less famous but equally fascinating is Auguste Bravais, a French naval officer who explored the Arctic. He discovered mathematical forms in both plants and crystals, only to succumb at an early age to an acute form of despair.

Turing, known for his breakthroughs in early computer science, did research on phyllotaxis. From a young age, he was drawn to "watching the daisies grow," seeking to understand why Fibonacci numbers occur in them. Investigating phyllotaxis was Turing's last, unfinished work, when he committed suicide under somewhat mysterious circumstances at 41.

As you might imagine, the act of *looking* is crucial to the study of phyllotaxis.[1] (The term, by the way, comes from the Ancient Greek words for "leaf arrangement.") Observing became a kind of extreme sport for some of the scientists in this book, who had superhuman powers of patience. Alexander Braun, a German botanist, examined the scales on hundreds of pinecones to prove his hypotheses on spirals. The drawings of pinecones that his sister made for him—some with every single scale hand labeled—are exquisite. As for Bravais, the naval officer, he and his brother painstakingly recorded the leaf angles observed in dozens of different wildflowers, from prickly teasels to yellow asphodels. As you will see, their delicate illustrations are equally dazzling.

Over time, the nature of scientists' phyllotaxis observations changed dramatically. As technology improved, scientists in the late nineteenth century began seeing plants not as static objects but as dynamical systems. They saw that as plants grow, their new leaves and petals do not fit predetermined patterns but instead develop where there is room for them, given limited options for placement. Understanding why this occurs—on a biological and physical level—is eye-opening.

This book intends to introduce you to some math (as far as you dare go), to offer you a peek into scientists' lives, and to give you a sense of how science works over the course of generations. Sometimes we assume that science proceeds in a neat linear way, always progressing toward Truth. But in reality, science is often messy, and scientists sometimes squabble among themselves and forget what has already been discovered long before. Plants display self-repeating patterns, and science, too, can be self-repeating.

Finally, it's worth mentioning that phyllotaxis is science, not mysticism. Plants don't grow in spirals because there is a goddess who loves spirals, or because aliens brought the golden angle to Earth. Phyllotaxis is based on hard data, and it can get technical—but it can also knock your socks off. The quest to understand why Fibonacci numbers appear in plants is an eminently satisfying one, located at the intersection of math, physics, and biology.

Now, here is where knowing a bit of French comes in handy. (Note that three of the coauthors are francophones: two from France and one from Québec.) Sometimes, French has a word that English simply cannot capture. *Le merveilleux* is one of those terms. Any attempt at an English translation sounds awful. The marvelous? The wondrous? Blech.

Perhaps French is better at conveying emotions than English, and an encounter with *le merveilleux* is always very moving. You bump into something that seems unbelievable yet is entirely real, something that opens your eyes to patterns you never saw before. There is pleasure in discovery, and you never see the world the same way again.

## Musings on Math

At first glance, no one would connect plants to math. Plants are from the everyday world all around us. We love seeing them on our walks through gardens, fields, and forests, and we even bring them into the strange habitat of our homes. There, if we remember to water them, they cheer us up. So we love plants, but—perhaps just as with the people we love and live with—we don't really look at them.

Math, by contrast, appears to be from another world. Some say it is a pure creation of our brains. Others argue that math has an actual existence, somewhere out there in an abstract world.

So how exactly do we connect plants with math? Math gets pulled down to our everyday world because of its usefulness. And this usefulness is deeply imprinted in our brains. Many animals have a rudimentary sense of numbers and magnitude. Scientists have detected the neurons used by mud-brown frogs to count the "chuck" sounds in their mating calls. The male with more chucks wins over the female. Ravens can match objects having the same area and perimeter.[2]

Other examples abound. In humans, this evolutionarily useful sense of basic numbers and magnitude then led to developing more sophisticated values for trade, the measuring of surfaces and volumes, the computing of taxes, and so on. Math can also send people into space. All this is very useful.

But applied to plants? Well, of course we can use math for such practical tasks as predicting or measuring the size of a crop. Yet we can also go much further, applying math to living matter in ways that have nothing to do with conventional ideas of usefulness. That's where the beauty of both math and plants unfurls before our very eyes.

Let's take a look:

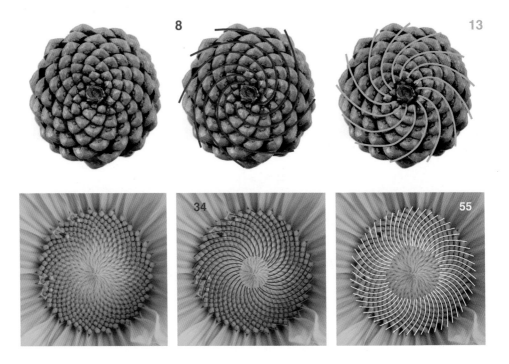

**FIG. 0.1** Spirals turning two directions, in pinecones and sunflowers.

On the pinecone and sunflower shown here, we have drawn spirals in two directions. We'll even count them for you: on the pinecone, there are 13 spirals turning in one direction and 8 in the other; and on the sunflower there are 34 and 55 spirals. These are precise numbers, all of them belonging to the Fibonacci sequence. (Recall that it begins 1, 1, 2, 3, 5, 8, 13, 21, . . .) We will see that this is quite a general property, not special to just the two examples shown here.

**So why do these numbers appear?** They don't do anything useful for us, and in fact we barely notice them. Are these numbers useful in some way for plants?

And why do plants display these particular mathematical numbers? Are they doing math? Were they doing math before humans began doing it? Did plants help us figure out math? These are the questions we are going to explore in this book, along with stories of the discoveries and the discoverers.

## Phyllotaxis in 10 Terms

This book is mostly about spirals. Before you can analyze them, however, you need to learn how scientists see them.

Like all sciences, the study of plant patterns has a specialized vocabulary—and admittedly some of the words are tongue twisters. What follows is a brief introduction to how scientists talk about this fascinating topic at the intersection of biology, mathematics, and physics.

### 1. SPIRALS

If you're looking at, say, a pinecone, the *spirals* formed by the scales are easy to see. You've probably known about them ever since you were a kid. The aloe plant in figure 0.2 makes it very clear.

Spirals can appear in a variety of contexts in plants, whether in the scales on a pineapple, the leaves of an artichoke, or the florets at the center of a dahlia. Spirals can also appear in the placement of leaves along a stem. It's helpful to think of tracing the thread on a screw, imagining leaves sprouting at equal intervals.

Some scientists have explained Fibonacci phyllotaxis as a way for plants to maximize sun exposure on their leaves, with minimal overlap. The authors of this book find that argument unconvincing, however. Although the sun is rarely directly overhead in most places on Earth (but instead inclined according to the

**FIG. 0.2** This plant is known as the "spiral aloe" for obvious reasons.

season), most plants still grow straight upward. In addition, there are many plants whose leaves grow directly above one another—such as the whorled plants described below—yet they have survived natural selection and continue to thrive. And finally, leaves can easily move toward the light to get more sun, as you can see in any houseplant growing in a window.[3]

## 2. PARASTICHY NUMBERS

What you may not have noticed is that you can often see plant spirals turning in two directions.

Scientists call each spiral a *parastichy* (while we pronounce this PAIR-a-sticky, others say pa-RAS-ticky).[4]

Usually, scientists count the spirals in both directions and talk about them as a pair $(m, n)$. These are known as *parastichy numbers*. In this book, we will call the smaller number $m$ and the larger number $n$.

**FIG. 0.3** The parastichy numbers for this pinecone are (8, 13).

## 3. FIBONACCI SEQUENCE

In a beautiful example of how nature meets math, the number of spirals on plants often fits the *Fibonacci sequence*.

The sequence is named after one of its popularizers, an Italian mathematician born in 1170. He was proposing a playful way to describe reproduction in rabbits. As it turns out, the sequence has little bearing on rabbits, but it's quite powerful for plants.

The Fibonacci sequence begins with two pairs of rabbits, 1 and 1. These get added together: 2. Then you simply keep adding the last two numbers to generate the next one: (1, 1, 2, 3, 5, 8, 13, 21, 34, 55, . . .).

Although the sequence is infinite, in the plant world it rarely appears beyond 144. The exception is a large sunflower head, which might have 233 spirals.

Note that in the pinecone in figure 0.3, both parastichy numbers—8 and 13—are Fibonacci numbers.

**4. DIVERGENCE ANGLE**

Scientists are also very interested in the angle between two consecutive leaves or petals (meaning that one grew after the next). This is called the *divergence angle*.

When plants display spiral phyllotaxis, one leaf does not grow directly above the next. Instead, the divergence angle between two leaves is often close to 137.5°. This is famously known as the golden angle.

Remarkably enough, as we will see, the golden angle is closely connected to the Fibonacci sequence.

**FIG. 0.4** Leaves in this *Aeonium* are numbered according to age. The divergence angle between leaf 8 and leaf 9 is shown in black, and that between leaf 5 and leaf 6 in white. Both are close to 137.5°.

## 5. GENERATIVE SPIRALS

Another useful tool imagines a spiral joining consecutive leaves along the length of a plant stem. This is called a *generative spiral*.

The beauty of the generative spiral is that it makes it easy to determine the average divergence angle between two leaves. To do this, first find two leaves growing almost directly above each other. (See the blue leaves indicated in fig. 0.5.) Then follow the generative spiral linking one leaf to the next. Count the number of leaves between the lower leaf and the upper one—but don't count the leaf you started with. Now count the number of full turns you made around the stem. The ratio of these two numbers gives you the average divergence angle.

Try this using the plant drawing in figure 0.5 at left, traveling between the two blue leaves:

$$\# \text{ turns} / \# \text{ new leaves} = 5 \text{ turns} / 13 = 0.385 \text{ turns} = 138.5°$$

**FIG. 0.5** At left, the dotted line indicates the generative spiral. At right, this spiral is shown by a white line winding around the plant. Note that the leaves marked in blue are almost perfectly aligned.

Note that this angle is close to the golden angle of 137.5°. (We will often return to this famous angle, starting in chapter 4 on the Bravais brothers.)

Now look at the second example, figure 0.5, right. Note that the leaves are numbered according to age, with older leaf 0 located almost directly below younger leaf 5. Following the white generative spiral, you make about two full turns to travel from one leaf to the other. And so, in this case, the average divergence angle is roughly

$$\text{\# turns} / \text{\# new leaves} = 2/5 \text{ turn} = 2/5 \cdot 360° = 144°$$

 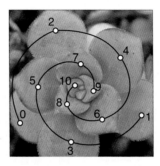 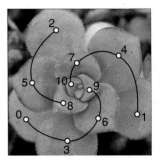

**FIG. 0.6** Three spirals are shown on the same *Aeonium*: at left, the white generative spiral passes through each leaf, generation by generation. The middle and right-hand photos show 2 counterclockwise parastichies and 3 clockwise parastichies, so the plant's parastichy numbers are (2, 3).

## 6. LATTICES

In addition to photographing and labeling actual plants, scientists have developed systems for modeling their geometry. Among the most useful are *lattices*.

Lattices show each leaf (or other plant organ) as a disk, in order to more clearly demonstrate plant patterns. (Well, technically the earliest lattices in botany used points instead of disks.)

In figure 0.7, the left-hand image is a *spiral lattice*, which views the plant from the top down.

The right-hand image is a *cylinder lattice*, which views the plant stem as if rolled out flat.

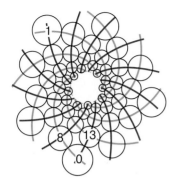
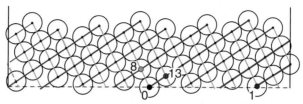

**FIG. 0.7** Two types of lattice models used by scientists to show plant geometry.

## 7. PRIMORDIA

Observing the formation of plant spirals, scientists study new plant organs as they emerge. When a new organ first emerges as a bump that will become a leaf, petal, scale, or other part, it's known as a *primordium*. The plural is *primordia*.

By using the word "primordia," scientists can speak about new plant organs in general, rather than being limited to particulars.

## 8. MERISTEMS

Given that most plants grow in spirals, how exactly does this occur? Scientists look closely at the very tip of the growing stem, around which the primordia appear.

The plant's tiny "organ factory" is called the *shoot apical meristem*. Scientists sometimes refer to this as the SAM.

Usually invisible to the naked eye, a SAM is best seen through a scanning electron microscope, or SEM. It's here that cell division takes place. Figure 0.8 is a SAM viewed by a SEM. The entire image of the meristem is about 1 mm (1/24 of an inch) and would fit on a pinhead.

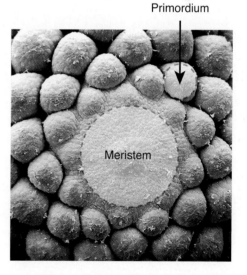

**FIG. 0.8** Scanning electron microscope (SEM) image of the shoot apical meristem of a spruce branch, surrounded by needle primordia.

**FIG. 0.9** Whorled phyllotaxis in a great horsetail (*Equisetum telmateia*).

## 9. DYNAMICAL SYSTEMS

For centuries, scientists saw phyllotaxis as static. They observed dried plants, preserved at a single moment of their existence.

Only in recent decades has phyllotaxis been understood as a *dynamical system*, a mathematical model that describes movement over time. Plant spirals are not static but instead emerge as a function of cell division and plant growth.

## 10. WHORLED PHYLLOTAXIS

While the majority of plants in nature exhibit spiral phyllotaxis, other growth patterns do exist. Among them, the most common is called *whorled*.

In whorled phyllotaxis, multiple organs grow at each level of the stem. In the ancient plants called horsetails, as many as 20 leaves can be found on a single level.

## ONE MORE . . .

One additional type of phyllotaxis, studied in detail by this book's authors, is the *quasi-symmetric phyllotaxis* found in such plants as strawberries and corn. Although these display some regularity that can be analyzed mathematically, the patterns are not as predictable as in plants displaying classic Fibonacci phyllotaxis.

# Try Your Hand

### Draw Smartphone Spirals

1. Take a photo with your smartphone of a flower whose spirals are very clear. If you're lucky, you might find this out in nature. Or else in a garden shop. Alternatively, take a photo of a flower in this book, like the dahlia in figure 0.10.
2. Using the edit function, click on the markup (or drawing) function. Choose a bright color for contrast and use your finger to trace all the clockwise spirals. Tip: don't try to draw spirals across the whole flower, only where they are most obvious—usually toward the outside. The center can get messy.
3. Take a screenshot.
4. Revert to the original photo. Now, using a different bright color, trace all the counterclockwise spirals. (In the dahlia, the counterclockwise ones are harder to see.)

**FIG. 0.10** A dahlia, useful for drawing and numbering spirals.

Did you end up with $(m, n)$ parastichies that are Fibonacci numbers? Hopefully *oui*! (This technique also works with pinecones.)

### Pin Pineapples

Pineapples display spirals running in two directions, often with a clear dominant side (i.e., some pineapples are left-handed, and some right-handed). Using colored pushpins is the easiest way to see and count the spirals. You may reach dead ends, where the spirals divide. These are "transitions," which we will explore in later chapters.

**FIG. 0.11** Colored pushpins make it easy to count spirals on a pineapple.

**Roll Your Own Cone (3D to 2D)**

In this activity, you will "unroll" a plant cone on dough. This is a good way to observe and count the plant spirals that wrap all the way around the cylinder.

1. Unrolling Your Cone
   a. Find a pinecone or other cone that's fairly symmetric. Here, we have used a nice closed cone from a black pine (*Pinus nigra*). (You can make your cone close up by leaving it in a humid environment.) Your cone should show clear, tight spirals and be firm enough to leave an imprint when rolled.
   b. Roll out some dough or modeling clay using a rolling pin (or a bottle). Your slab should be about the thickness of pie dough. Make it a little wider than your cone and with a length about eight times its diameter.
   c. On your cone, mark a point midway between the top and bottom, using a blob of paint or nail polish. This will mark one full turn of your cone.
   d. Roll out your cone onto the dough, using the fingers on both hands to go as straight as possible. Apply firm and constant pressure. Roll at least two full turns of the pinecone.
   e. Check to make sure your paint mark shows up for each full turn.

2. Tracing and Counting Parastichies
   a. Choose a pair of consecutive marks on your imprint.
   b. With your finger, trace the parastichy that passes through the left-hand mark and slants upward, following the lines between the scales. This will be 1.
   c. Next, trace and count the parastichies that run parallel to the first one.
   d. Stop counting when you reach the parastichy *before* the one that passes through the right-hand mark. We got 5 for our pinecone.
   e. Repeat the process, this time using the parastichies slanting downward. We got 8 for our pinecone.
   f. Sometimes your cone, like ours in the picture, will show transitions—places where some parastichies split or come to a dead end. First, focus on the part of the slab with a regular pattern.

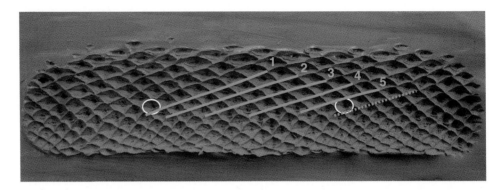

**FIG. 0.12** Tracing the first set of parastichies.

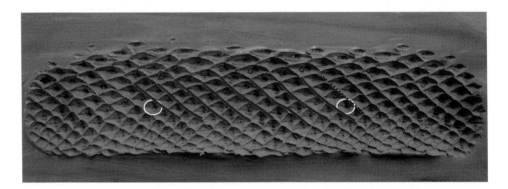

**FIG. 0.13** Tracing the second set of parastichies.

**FIG. 0.14** Counting the steps reveals the parastichy numbers.

g. Can you also count the steeper, upward-slanting parastichies in the lower part of our cone? (It should come as a nice surprise). Do you notice any change in the number of downward parastichies?

h. Another approach is to trace one parastichy going up from a chosen point and another going down through the same point one full turn later. (These are the same points on the pinecone, meeting when you complete the turn.) If your imprint is wide enough, these lines will cross (shown in fig. 0.14 as a black dot), forming a triangle. Count the number of steps going up to that intersection and then the number going down. In the photo, we go up 8 (green) steps. At each of these steps, we cross one red, downward parastichy. So the number of steps on the *green* line equals the number of *red* parastichies. Likewise, going down the *red* line, you can count 5 steps, each crossing a *green* parastichy. So the number of green, upward parastichies will be 5.

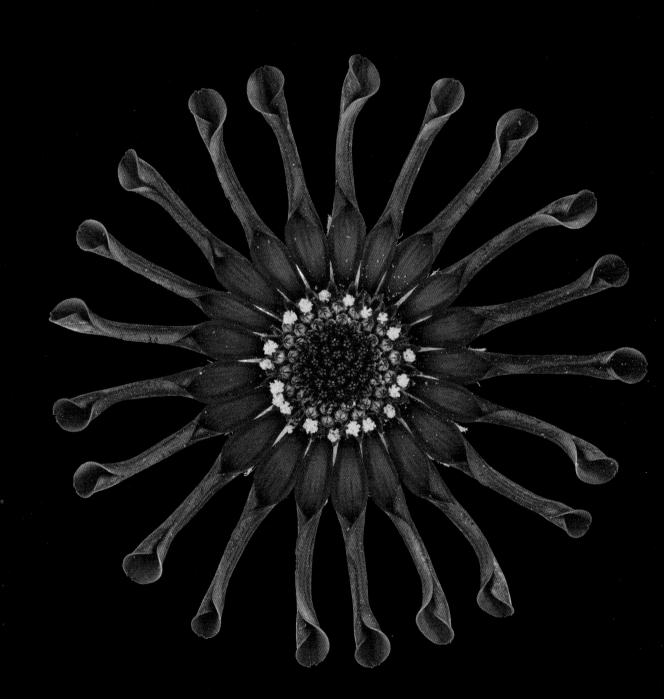

PART I

# Who Noticed First?

CHAPTER 1

# The (So-Called) Fibonacci Sequence in History

**A Fibonacci Poem**[1]
One
one
then two,
next three, five,
eight—add the last two
and keep going infinitely.
Call this sequence Fibonacci? It's much more ancient!

For thousands of years, humans have been keen observers of plant life, as a matter of survival. They relied on plants for sustenance and medicine, and at least as far back as the Ancient Egyptians, they used plant motifs in their art and architecture. When exactly they began observing plant spirals and the mathematical order within plants is a matter of speculation. We can know only what has come down to us in writing or in recovered artifacts.

The Ancient Egyptians used the forms of palm, papyrus, and other plants in their architectural columns. In hieroglyphics, the symbol for 1,000 is a lotus plant—however, it is drawn in a stylized way and has no relation to the anatomy of a lotus plant.[2] The Ancient Egyptian system for representing large numbers was quite simple: 1,000 was one lotus; 2,000 was two lotuses.

FIG. 1.1 The Ancient Egyptians wrote the number 4,622 in this way, using a lotus blossom (at top) to represent 1,000. Can you guess the symbols for hundreds, tens, and ones?

The study of phyllotaxis got off the ground some 2,300 years ago in Ancient Greece, when the philosopher Theophrastus (370–285 BCE) wrote down his general observations of leaf and branch arrangements. In his *Enquiry into Plants*, he noted that plants "that have flat leaves have them in a regular series," offering myrtle as an example.[3] (We would now describe the myrtle as whorled by two, with two leaves at the same height and the next pair crosswise.) He also observed that "the branches of the silver-fir are arranged opposite one another; and in some cases the branches are at equal distances apart, and correspond in number, as where they are in three rows."[4]

The next milestone dates from about 135 BCE, when the Chinese philosopher Han Ying (ca. 200–ca. 120 BCE) made an explicit link between plants and numbers, in his treatise on moral discourses:

> Flowers of plants and trees are generally five-pointed, but those of snow, which are called *ying*, are always six-pointed.[5]

As Philip Ball points out in his book *Nature's Patterns*, "It is a casual reference, as though he is mentioning something that everyone already knew."[6] (Much later, in the sixteenth century, the astronomer Johannes Kepler would make observations very similar to those of Han Ying, as we will see.)

Then came the prolific Roman writer Pliny the Elder (ca. 23–79 CE), who died in Pompeii during the eruption of Mount Vesuvius. In his *Natural History*, Pliny counted the leaves of the aparine plant and observed its growth pattern, describing

**FIGURE 1.2** Theophrastus noted the regular arrangement of leaves in the myrtle plant. Can you see it?

it as a "ramose, hairy plant with five or six leaves at regular intervals, arranged circularly around the branches."[7] The plant has been identified more precisely as *Galium aparine,* whose seeds are burs that stick to animal fur and human passersby (the name *aparine* means "clinging"). But if this is the right plant, its leaves more commonly grow in whorls of six and eight, and only rarely five.[8]

## Leonardo de Pisa, a.k.a. Fibonacci

Recall that the numbers playing a starring role in plant patterns are those of the Fibonacci sequence:

1, 1, 2, 3, 5, 8, 13, . . .

in which each number is the sum of the previous two. The sequence was repeatedly "discovered" by scientists and even Sanskrit poets with mathematical inclinations. Not until much later, however, would it be linked to plants.

In Europe, this wonderfully pleasing number sequence made its debut in 1202. That was the year that Leonardo de Pisa (ca. 1170–after 1240) completed a massive work that helped bring a mathematical revolution to Italy and then the entire Western world. The author is known today as Fibonacci, the nickname he was given in the nineteenth century, for he described himself as *filius Bonacci* in his book. (The meaning of this phrase is debated: was he a son of the Bonacci family, or was Bonacci his father's nickname as a "good guy"?) For the sake of simplicity, we'll call him Fibonacci in this chapter.

Fibonacci's real breakthrough was showing Europeans the beauty and power of Indo-Arabic numerals. Before Fibonacci spread the word, Italian merchants were stuck with Roman numerals. Just try doing MCCII times XXXIV! If merchants needed more complex calculations, they had to find someone expert in the use of the local abacus, which consisted of pebbles on a board marked with lines.

Although Fibonacci was born in Pisa, he spent much of his childhood in North Africa. His father was a successful merchant and customs official in Bugia—now Béjaïa, Algeria—for the Italians doing business there. As Fibonacci recalled, his father "summoned me to him while I was still a child, and having an eye to usefulness and future convenience, desired me to stay there and receive instruction in the school of accounting."[9]

**The Book of Calculations**

After traveling the region, soaking up all the mathematics he could, Fibonacci sat down to write his magnum opus, the *Liber abaci*. Careless translators have called this the "Book of the Abacus," but if anything, the book is about doing computations *without* an abacus.[10] Fibonacci wrote by hand—the printing press hadn't been invented yet—but when printed hundreds of years later, the book ran to 468 pages. The original volume was written in Latin, the language of Western scholars, and so had a somewhat limited audience. But Fibonacci then followed up with a shorter and more practical version in Italian, which was much copied and imitated. Together, these works launched a revolution in mathematics and accounting, first in Italy and then in the wider world.

Essentially, Fibonacci was the Steve Jobs of the thirteenth century, according to British mathematician Keith Devlin, author of *The Man of Numbers: Fibonacci's Arithmetic Revolution*. What Fibonacci did, Devlin wrote, was "every bit as revolutionary as the personal computer pioneers who in the 1980s . . . made computers available to, and usable by, anyone."[11]

A running theme in *Do Plants Know Math?* is that science often spirals back upon itself, rather than moving forward in a straight line. Fibonacci's work is no exception. He wrote that he received

> marvellous instruction in the art of the nine Indian figures . . . [which] pleased me so much above all else, and I learned from . . . whoever was learned in it, from nearby Egypt, Syria, Greece, Sicily and Provence, and their various methods.[12]

Indeed, historians have traced many of the examples in *Liber abaci* to works by the Persian mathematician Muhammad ibn Mūsā al-Khwārizmī and the Egyptian mathematician Abū Kāmil, who lived more than two centuries before

Fibonacci.[13] (Note that al-Khwārizmī's name, Latinized as Algorithmi, gave rise to the word "algorithm.") But of course, it would have been impossible in the thirteenth century for Fibonacci to acknowledge that much of his book simply translated Islamic authors[14]—not at the time of the Crusades, when Christian European knights were trying to defeat Islam in Jerusalem.

Finally, it's important to note that Fibonacci was not just a talented translator and popularizer. Many scholars consider his *Liber quadratorum* (*The Book of Squares*) the first major European advance in number theory since the work of the Greek mathematician Diophantus a thousand years earlier.[15]

## The Rabbit Problem

Most of the scientists featured in this book are both lovers of abstraction and keen observers of nature. Fibonacci is a bit of an exception: the rabbit problem that laid out the famous sequence was a puzzle that had nothing to do with the reproduction of actual rabbits.

Here is the problem, roughly as stated by Fibonacci in his *Liber abaci*:[16]

> A person put a pair of rabbits inside a walled enclosure. How many pairs of rabbits will that pair produce in a year, assuming that every month each pair begets a new pair that can reproduce from the second month on?

Fibonacci goes on to solve the problem month by month. (Note that he assumes the first pair already bears kits the first month, in effect skipping the first 1 in the sequence. For a more strictly accurate approach that begins with 1, 1, 2, . . . , see fig. 1.3 .) He writes:

> Because the pair mentioned above bore in the first month, you will double the number: there will be two pairs. One of these, namely the first pair, bears in the second month, and thus in the second month there are 3 pairs; of these, in one month two will be pregnant, and in the third month 2 pairs of rabbits are born, and thus there are 5 pairs in the month.

He continues until he gets to the end of the 12th month and 377 pairs of rabbits. Not until he reaches the solution does he present the algorithm explicitly, explaining the table in the margin (see fig. 1.4) where the sequence appears:

You can indeed see in the margin how we proceeded, adding the first number to the second, namely the 1 to the 2, and the second to the third, and the third to the fourth, and the fourth to the fifth, and thus one after another until we added the tenth to the eleventh, namely the 144 to the 233, and we had the above written sum of rabbits, namely 377, and so you can find the answer in order for an unending number of months.

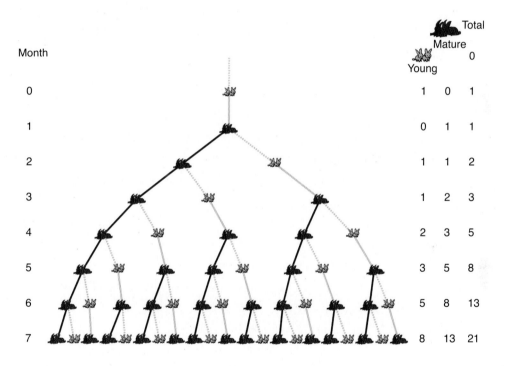

**FIG. 1.3** Illustration of the solution to Fibonacci's rabbit problem. The first pair arrives as babies at month 0 and reproduces in month 2, allowing the sequence to begin 1, 1, 2, . . . Do you see the parents that reproduce every month (red lines)? And their successive new babies (yellow lines, running in the opposite direction)?

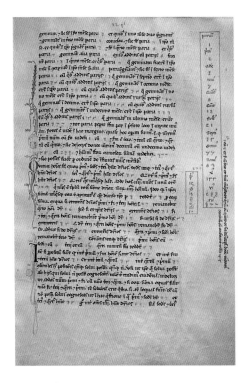

**FIG. 1.4** Fibonacci's solution to the rabbit problem in his *Liber abaci*. Month-by-month results appear in his table at far right, in the red box. (Note that the 4's look like r's, and the 3's resemble z's. What about the 5's?)

Not until the nineteenth century was this pattern named the "Fibonacci sequence," when French number theorist Édouard Lucas (1842–1891) pointed to Fibonacci as its discoverer.[17] But in fact the sequence had already been used for many centuries by Indian poets.

**The Sanskrit Story**

If we dig a bit deeper, it turns out that the so-called Fibonacci sequence and its recurrence rule were known in India to Sanskrit poets long before Fibonacci, arguably as early as around the third century BCE. Some believe that the sequence should be named the Hemachandra-Fibonacci sequence, based on the writings of the Indian mathematician-poet Hemachandra (हेमचन्द्र) (ca. 1088–ca. 1173). But Hemachandra was not the first: he noted that his work was based on that of the Indian mathematician Gopala, part of a chain that apparently goes back centuries![18] (This is a fine example of Stigler's law of eponymy, which states that no scientific discovery is named for its actual discoverer. We will see this many times over the course of this book.)

Still, the beauty of acknowledging Hemachandra, a Jain scholar in the court of King Kumarapala, is that he wrote very clearly on the topic and stated the sequence in numbers. For him, the problem had nothing to do with rabbits—or plants—but instead described the possibilities for composing Sanskrit poetry using different numbers of long and short syllables. Poetry was extremely important to the culture at the time, used as a way to remember discoveries of all types, including mathematical and scientific ones.

Hemachandra completed his treatise on poetic meters in about 1150, some 50 years before Fibonacci finished writing *Liber abaci*. Hemachandra wrote, "The sum of the last and the last but one numbers [of variations] is [that] of the *mātrāvrttas* coming next."[19] The number of possible *mātrāvrttas*—the meters known as *mātrā* that use specific counts of long and short syllables—generates the sequence. He then commented on his rule: "Thus—1, 2, 3, 5, 8, 13, 21, 34 and so on."[20]

## Syllables in Sanskrit

In Sanskrit poetry, there are two distinct types of syllables: short, one-beat syllables; and long, two-beat syllables. (This is a little like Morse code, with its dots and dashes.) A natural question is:

> How many different combinations of short and long syllables can a poet use in a verse with a given number of beats?

To find the answer, it's easiest to start with the shortest verses.

> For a one-beat verse, there is just one option: one short syllable.

> For a verse with two beats, you can use either one long syllable or two short ones. So there are two possible combinations.

> For a verse with three beats, you can use either three short syllables, or one short and one long syllable. But for the latter, there are two ways to arrange them: short-long or long-short, making a total of three possible combinations.

Now let's present the information more concisely, denoting the short syllables as 1, and the long ones as 2. The game is then the following:

> Given some integer $n$, write all possible numbers containing only the digits 1 and 2, in such a way that the sum of the digits is $n$. Table 1 shows the answers

for *n* up to 5, and invites you to fill in the blanks for greater values of *n*. Can you craft an argument for why these syllables follow the Fibonacci rule at any arbitrary step?

**TABLE 1 The possible verses of length n, using beats of one length or two.**

| *Verse length, in beats* | *Possible combinations* | *Number of combinations* |
|---|---|---|
| 1 | 1 | 1 |
| 2 | 11, 2 | 2 |
| 3 | 111, 12, 21 | 3 |
| 4 | 1111, 211, 121, 112, 22 | 5 |
| 5 | 11111, 2111, 1211, 1121, 1112, 221, 212, 122 | 8 |
| [6] | ? | ? |
| [7] | ? | ? |

Indian prosody experts had, as early as 400 BCE, figured out that you could guess the number of possible combinations in the next row based on the previous two rows.[21]

At this point in history, we have chalked up the first plant observations and the discovery of so-called Fibonacci numbers. But the plant descriptions were still rudimentary, and the Fibonacci numbers were not connected to plants but rather to Sanskrit poetry or hypothetical rabbits. The next leap in the science of phyllotaxis would not occur until the Renaissance, through the observations of none other than Leonardo da Vinci.

# Try Your Hand

## See Fibonacci in Honeybees

Although the Fibonacci sequence does not actually predict rabbit reproduction, it perfectly describes the genealogy of male drone honeybees! Here's how it works:

- The male drone has just one parent, the queen bee.
- It has two grandparents, because its mother had two parents, a male and a female.
- It has three great-grandparents: its grandmother had two parents, but its grandfather had only one.

Can you make a genealogical chart showing a bee drone's ancestors for five generations?

## Draw Your Own Fibonacci Spiral

1. Print out a piece of graph paper with eight squares to the inch and turn it so the long side is horizontal.
2. Using a sharp pencil, outline one square located toward the right lower part of the page.
3. Outline the next square to its left.
4. Above these two squares, draw a $2 \times 2$ box. Rotate the page 90 degrees to the right.
5. Draw a $3 \times 3$ box above the two previous squares. Rotate the page 90 degrees to the right.
6. Draw a $5 \times 5$ box above the two previous squares. Keep going until you have made a $21 \times 21$ box. One edge of each box will be the sum of the two boxes before it, matching the Fibonacci sequence.
7. Finally, draw quarter-circle arcs inside each square, using a compass if you have one. The radius of each arc should match the length of the side of the box. Draw the arcs so that they form a continuous counterclockwise spiral,

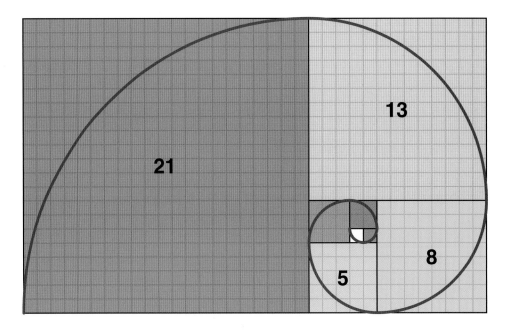

**FIG. 1.5** A Fibonacci spiral on graph paper.

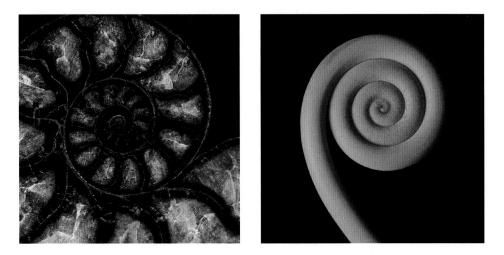

**FIG. 1.6** The logarithmic spiral in the fossil ammonite at left is often confused with a Fibonacci spiral. And while the spiral in the fern at right starts out looking Fibonacci-esque, as the stem's thickness stabilizes, it approaches an Archimedean spiral.

ending in the bottom left corner of the 21×21 square. If you'd like, color each box a different color.

**Note:** Often, one sees a nautilus shell pictured next to this Fibonacci spiral (or a similar spiral with golden-proportion rectangles). Although the shapes are similar, a nautilus shell is actually a logarithmic spiral, whose radius increases exponentially with the angle. If you like math questions, ponder why this is not (perfectly) the case for a Fibonacci spiral.

More generally, while many types of spirals occur in nature, it is important to recognize that they may result from very different growth mechanisms. The way a nautilus shell develops has little in common with plants. Similarly, the fiddlehead of a fern follows an entirely different growth mechanism from Fibonacci phyllotaxis.

CHAPTER 2

# Plant Patterns in Leonardo's Notebook

In
his
mirror
writing, he
described four modes of
leaf arrangement. This was a first.
But Leonardo crossed one out. Who can crack his code?

Using his extraordinary powers of observation, Leonardo da Vinci[1] (1452–1519) turned his gaze to the botanical world. Not surprisingly, he was, to our knowledge, the first person to record specific patterns in plants. Before scientific botany had even gotten off the ground, he drew and painted plants with exacting precision. And he demanded the same from his students, exhorting: "Painter, know that you cannot be good unless you are a universal master whose art represents every kind of form produced in nature."[2]

Coming into the world, Leonardo had few advantages, as his parents were not married. His father was a notary, and although little is known about his mother, recent scholarship suggests she may have been an enslaved woman, possibly from Central Asia. Being born out of wedlock freed Leonardo to be creative, rather than follow in his father's footsteps as a notary. With little formal education, at age 14 Leonardo joined the Florence workshop of Andrea del Verrocchio, where he learned painting and sculpting. At 20, he began working as an independent painter nurtured by the Medici family, powerful Florentine bankers and patrons of the arts. He quickly rose to fame as a painter, philosopher, and inventor, employed by a series of important patrons. Among his other roles, he served for a time as a military architect for Cesare Borgia, the noble who inspired Machiavelli's *The Prince*. At the end of his long life, Leonardo was living near King François I in France, with the *Mona Lisa* in his room.

Among Leonardo's most astounding legacies are his notebooks: 13,000 pages of sketches and notes on art, engineering, science, and nature, written mostly in his mirror-image cursive. Although he intended to organize his thoughts into formal treatises, he never did.[3] This was a loss for science, as his findings show amazing breadth and depth, in many cases preceding published discoveries by centuries—in botany, too. In just half a page, Leonardo set down a wholly original and groundbreaking classification of plant patterns.

In recent years, Leonardo's reputation for botanical exactness has spawned an art history controversy. In the London version of his masterpiece *Virgin of the Rocks*, experts noticed that the daffodil leaves are depicted inaccurately. Some art historians have argued that therefore the entire painting must be a fake. (The controversy has never been entirely resolved.)

### Decoding Leonardo's Classifications

To our knowledge, Leonardo was the first person to classify plant patterns, describing specific ways that leaves or branches are inserted around a stem. Although he writes on this notebook page that there are four "modes" of leaf insertion, he describes only three. But a tantalizing scribble indicates his sketch of a fourth mode and a few crossed-out words that have not been transcribed before. (Below, you'll have a chance to play da Vinci decoder.)

His notebook passage on phyllotaxis begins at the top of the page.

**FIG. 2.1** The page on phyllotaxis from Leonardo's Notebook G, flipped to make his mirror writing more legible.

Leonardo's spelling in Italian was somewhat eccentric, but we have transcribed the text exactly as he wrote it, so that you too can try deciphering his handwriting. (Don't worry, there's an English translation below.) Note that a vowel with a bar above it is a contraction of that vowel followed by *n*. So *ā* = *an*. He also uses a *p* with a large curl to signify *pr*, as in *sopra*.

**Del nasscimēto de' rami nelle piāte**

Tale he il nasscimēto delle ramifichationi delle
piāte sopra i lor rami prīcipali quele quella del
nasscimēto delle foglie sopra li ramichuli del
medessimo āno desse foglie le quali foglie anno
quatro modi di proccedere luna piu alta chellaltro
eprimo piu vniuersale he chessenpre la sesta di
sopra nassce sopra la sessta di sotto e il secondo
heche le 2 terze di sopra sō sopra le due terze di
sotto el terzo modo he chella 3ª di
sopra essopra la 3ª di sotto [al 4 . . . ?]

**Of the Insertion of the Branches on Plants**

Such as the growth of the ramification of
plants is on their principal branches,
so is that of the leaves on the shoots
of the same plant. These leaves have
four modes of growing one above another.
The first, which is the most universal, is that
the sixth always originates over the sixth below;
the second is that two third ones above are over
the two third ones below; and the third way is that
the third above is over the third below. (*crossed out:* the 4th . . . ?)[4]

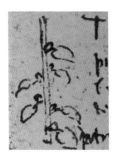

**FIG. 2.2** The first mode. Do you see the sixth leaf directly above the first?

**FIG. 2.3** The second mode described by Leonardo. Pairs of leaves appear at the same height. Can you see the rotation of successive pairs?

**FIG. 2.4** The third mode. Do you see how every other leaf is aligned?

These are extraordinary observations for their time. Leonardo recognizes that the side branches display the same types of arrangements as the leaves on a stem.[5] He then claims that for the leaves, there are four modes, or types of arrangement. We will now unpack his very succinct phrases to see what he has observed in more detail.

The first mode Leonardo describes will later be known as a quincunx. As he correctly notes, this arrangement is often found in nature (see fig. 0.5, right). Five leaves (*quin-*) form the pattern, with the sixth leaf aligned directly above the first. Surprisingly, here Leonardo's drawing is less than illuminating. But in a related section of the notebooks, Leonardo describes this mode again, writing that "the insertion of the leaves *turns round each branch* in such a way, as that the sixth leaf above is produced over the sixth leaf below."[6] This is the first hint at spirals in plants that we know of.[7]

In the second mode described by Leonardo, the leaves appear in pairs at the same level, pointing away from each other, and the next pair forms a cross with the previous one. The third pair is aligned with the first. This order will later be called decussate, whorled by two, or (2, 2) phyllotaxis.

In the third mode, the leaves appear one at a time, pointing in the direction opposite to the previous one, so that the third leaf is aligned with the first. This mode, which came to be called distichous, or (1, 1) phyllotaxis, is common in grasses.

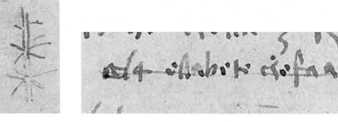

**FIG. 2.5** What about the mysterious fourth mode, quickly sketched at left? Leonardo has crossed out the beginning of a description that the manuscript's transcriber, Jean Paul Richter, does not mention.

To the right of the third mode, just after the crossed-out sentence, Leonardo has roughly sketched a stem that seems to have four branches growing at the same level. This pattern appears to repeat in the (very faint) bottom section. The top two branches, however, seem to break the pattern.

### What Was Crossed Out?

After intensely studying the mysterious crossed-out phrase, Chris and his daughter, Marguerite, came up with a possible interpretation. They matched Leonardo's handwriting and his idiosyncratic spelling to the texts previously deciphered by Richter. Here is their best contender:

> al 4 el'abete che fa [illegible] = For the 4th [mode], it is the fir that makes [illegible]

If that's correct, then Leonardo was thinking of a fir tree's branches as an example of his fourth mode. In nature, these grow in groups of four (or sometimes more), with single branches sometimes appearing in between the groups. Although the groups of branches may all appear to be at the same height, they are actually not quite level. In fact, if you look at a small fir (e.g., a Christmas tree), you will see that the needles growing out of the trunk together with young branches form a pretty dense set of very regular spirals, with none of the nodes at exactly the same height.

Another possibility is that Leonardo was trying to describe "whorled" arrangements, in which some number of leaves (sometimes four) grow at the same level, spread evenly around

**FIG. 2.6** A Douglas fir (*Pseudotsuga menziesii*).

PLANT PATTERNS IN LEONARDO'S NOTEBOOK

the stem. At the next node, the same number of leaves appears, but with the leaves above pointing midway between the two below. But because Leonardo's drawing is unclear, we're only speculating here. (For more on Leonardo's insights into trees and vasculature, see the appendix.)

With his razor-sharp observations, Leonardo made important contributions to phyllotaxis during the Renaissance.[8] He was the first to describe several types of leaf arrangements in plants, and he introduced a spiral form linked to the number 5, a Fibonacci number. Yet Leonardo stopped short of making a real connection between Fibonacci numbers and plants. The German astronomer Johannes Kepler will take us another step along that path, linking the Fibonacci sequence to another fascinating concept, the golden ratio.

# CHAPTER 3

# The Golden Ratio as a New Year's Gift

> Five
> was
> Kepler's
> most adored
> number. He saw that
> many flowers have five petals.
> And the golden ratio? From pentagons, of course.

Coming a century after Leonardo da Vinci, the story that relates the German astronomer Johannes Kepler (1571–1630) to phyllotaxis is undeniably an odd one. He made a notable leap forward, being the first to place three key elements of phyllotaxis on a single page: five-petaled flowers, the golden ratio, and the Fibonacci sequence. But he never synthesized these observations into a study of plant spirals, instead taking off in a much wackier direction.

Perhaps he was drawn to plant forms because his mother, Katharina Guldenmann Kepler, was an herbal healer, a practice she learned from her aunt. This turned out to be a terrible occupation: during the religious turmoil of the Reformation, both women were accused of being witches. The aunt was burned at the stake. Kepler's mother was first accused of sorcery in 1615, after a customer claimed that a magical potion caused her to suffer excruciating pain. Katharina

**FIG. 3.1** Johannes Kepler, royal astronomer and astrologer, at left. His mother, Katharina Guldenmann Kepler, at right, was accused of sorcery.

Kepler was ultimately arrested in 1620 and spent 14 months in prison, chained to the floor. Her son dropped his scientific work to prepare her legal defense, using his influence as royal astrologer to get her released. She died the following year.

As both a scientist and a Christian theologian, Johannes Kepler is often seen as a transitional figure between the Renaissance and the modern scientific era. He was obsessed with perfect forms and proportions, seeing in them the hand of God. "Geometry has two great treasures," he wrote in 1596. "One is the relationship of the hypotenuse to the sides in a right-angled triangle [the Pythagorean theorem], the other the division of a line into extreme and mean ratio [the golden ratio]."[1] He called the ratio "a jewel." And Kepler actually wept when, in a eureka moment, he thought he had discovered a relationship between Platonic solids and planetary orbits![2] (For more details, see the appendix.)

Yet in his search to prove godly perfection, he was also hungry for the best scientific data of his age. Kepler had worked for several years, somewhat unhappily, under the Danish astronomer Tycho Brahe. When Brahe died, Kepler was given his trove of extraordinarily precise astronomical measurements. Using these data, Kepler made huge and lasting discoveries. He developed his three laws of planetary motion, the first law being that the planets orbit the sun along elliptical paths. Although he believed this complex motion to be true, he never quite relinquished his simpler (and incompatible) Platonic ideal of circular orbits—somehow finding room for both inside his head.

**Six-Pointed Snowflakes, Five-Petaled Flowers**

For the new year 1611, Kepler wrote a short gift book for his friend and benefactor, the royal adviser Johannes Wacker von Wackenfels. Composed in Latin, *A New Year's Gift: The Six-Cornered Snowflake* is playful but profound, with Kepler pondering patterns both organic and inorganic.[3] He mused on why snowflakes have six equal branches—an inquiry that launched the study of crystallography. He gazed at the seeds of pomegranates—leading him to devise a math problem regarding the optimal packing of spheres that went unsolved for four centuries. (For more details, see the appendix.) He then moved on to three tantalizing paragraphs that hint at phyllotaxis but don't pursue it, having linked plants to the golden mean and the Fibonacci sequence. He wrote the following (section names added):

### I. On plants

We may ask why all trees and bushes—or at least most of them—unfold a flower in a five-sided pattern, with five petals. In apple- and pear-trees this flower is followed by a fruit likewise divided into five or into the related number, ten. Inside there are always five compartments for the reception of the seeds, and ten veins. This is true of cucumbers and others of that kind. It is here, I insist, that a consideration of the beauty or special quality of the shape that has characterized the soul of these plants, would be in place, and I will incidentally divulge my thoughts on this subject.

**FIG. 3.2** Kepler observed that many flowers, especially those on fruit trees, have five petals. This example is the succulent *Stapelia grandiflora*.

**FIG. 3.3** A dodecahedron and an icosahedron, drawn by Leonardo da Vinci for his friend Luca Pacioli's treatise *De divina proportione* (The golden section). Following Kepler's lead, do you see the "three-angles" in the polyhedron made of pentagons, and the "five-angles" in the one made of triangles?

## II. On the golden section and Fibonacci

Of the two regular solids, the dodecahedron and the icosahedron, the former is made up precisely of pentagons, the latter of triangles but triangles that meet five at a point. Both of these solids, and indeed the structure of the pentagon itself, cannot be formed without the divine proportion [golden section], as modern geometers call it. It is so arranged that the two lesser terms of a progressive series together constitute the third, and the two last, when added, make the immediately subsequent term and so on to infinity, as the same proportion continues unbroken.

## III. On the Fibonacci sequence and convergence

It is impossible to provide a perfect example in round numbers. However, the further we advance from the number one, the more perfect the example becomes. Let the smallest numbers be 1 and 1, which you must imagine as

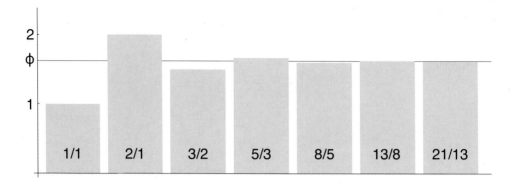

**FIG. 3.4** A visual approximation of the golden ratio, representing successive quotients of Fibonacci numbers. The values are 1, 2, 1.5, 1.666 . . . , 1.6, 1.625, 1.615 . . . , which slowly approach ϕ = 1.61803398874 . . .

unequal. Add them, and the sum will be 2; add to this the greater of the 1's, result 3; add 2 to this, and get 5; add 3, get 8; 5 to 8, 13; 8 to 13, 21. As 5 is to 8, so 8 is to 13, approximately, and as 8 to 13, so 13 is to 21, approximately.[4]

Did Kepler independently come up with the idea that ratios of Fibonacci numbers converge to the golden mean?[5] It's certainly possible, given that Kepler wrote about proportion in many variations, from planetary motion to musical intervals. But some scholars claim that other mathematicians saw this convergence first.[6]

Following his Fibonacci paragraph, Kepler's next sentence is a strange one, relating the golden ratio to sexual reproduction in plants. "I suspect that the generative faculty in seeds is formed in the likeness of this self-propagating proportion,"[7] he writes. He then sees the flower's pentagon as a sign of that same "generative faculty." Kepler's bizarre ideas on procreation and geometry involved such ideas as sex between two squares whose dimensions involve Fibonacci numbers![8] For better or worse, however, that path will not lead us anywhere on the road to phyllotaxis. The discovery of Fibonacci numbers in plants would have to wait another century, until Charles Bonnet began his experiments on plants and dew in Switzerland.

# Golden Mean and Golden Angle

## There are an infinite number of ways to cut a given segment into two.

For example, at one-eighth of the length we get the ratios Red/Gold = 7/1 and Whole/Red = 8/7:

Whole
Red  Gold

Rolling the lines into a circle, we get an angle of 45°:

At one-fourth of the length we get the ratios Red/Gold = 3/1 and Whole/Red = 4/3:

Whole
Red  Gold

Rolling the lines into a circle, we get an angle of 90°:

We get the Golden Mean, symbolized by φ, only when the two ratios Red/Gold and Whole/Red are equal:

Whole
Red  Gold

$$\phi = \frac{\text{Red}}{\text{Gold}} = \frac{\text{Whole}}{\text{Red}} = \frac{1 + \sqrt{5}}{2} = 1.61803...$$

Rolling the lines into a circle, we get the Golden Angle of ~137.5°:

## Ancient Construction of the Golden Mean

This ancient construction of the Golden Mean, using only a rule and a compass, appeared circa 300 B.C. in Euclid's Elements. It is used in the Elements to construct pentagrams (figure below), pentagons, and platonic solids.

**FIG. 3.5** An introduction to the golden ratio, from a Smith College exhibition on phyllotaxis.

**The Golden Ratio in Fact and Fiction**

**FACT**

There are many ways to cut a line segment into two pieces. But there is only one way to cut it such that the ratio of the whole to the longer part is equal to the ratio of the longer part to the shorter. This ratio is often denoted by the Greek letter $\phi$ (pronounced "fye"):

$$\phi = \frac{\sqrt{5}+1}{2} \approx 1.61803398874$$

Euclid knew the ratio, and he used it in his *Elements* to build pentagons, pentagrams, and Platonic solids like the dodecahedron. Kepler also knew $\phi$, and he saw that it could be obtained at any precision by taking quotients of Fibonacci numbers.

**FICTION**

If you look up the term "golden ratio," "golden mean," or "divine proportion" on the internet, you'll see that the proportion was supposedly used in the design of countless Greek monuments and Renaissance paintings. Closer examination reveals, however, that this is probably not true.

The origin of the myth is an interesting story in itself. All indications seem to point to a single book, published in 1854 by the German psychologist Adolph Zeising. Zeising's tome became a huge bestseller in Europe, despite its absurdly long title: *An Exposition of a New Theory of the Proportions of the Human Body, Based on a Previously Unrecognized Fundamental Morphological Law Which Permeates All of Nature and Art . . .* , etc.[9] This highly influential work, along with Zeising's many articles on the topic, helped spread the myth throughout Europe, often with a tone of underlying racism.

But then, what are we to make of the fact that Leonardo da Vinci illustrated Pacioli's *De divina proportione*, a Renaissance book that undeniably focused on the golden ratio and its role in the construction of geometric figures? Albert van der Schoot, a Dutch philosopher and author of a history of the golden ratio, explained that Pacioli "writes from a mathematical and metaphysical point of view (he was a Franciscan professor of mathematics), but *never* from a 'practical' point of view."[10] Van der Schoot underscored that the idea of "shaping something *according to* this proportion is a 19th-century idea,"[11] not a Renaissance one.

As for Leonardo's perspective, in his own writings he used the term "divine proportion" only three times, as Van der Schoot pointed out. In contrast to his friend Pacioli, whose book Leonardo illustrated, he never used the term to mean what we now call the "golden section." Instead, he meant "beautiful proportionality." Leonardo was a keen observer of all aspects of nature, including phyllotaxis, but he never referred to the golden section by any name.[12]

PART II

# Could Early Scientists Explain Plant Spirals?

CHAPTER 4

# First Spirals in the Dew

He
saw
spirals:
Bonnet was
first to name five types.
Until his eyesight fizzled out,
his observations were sharp, not hallucinations.

Now for the first time, we will see plant spirals appear before scientists' eyes. But the discovery was made while trying to solve a different problem. In the 1740s, two scientists of the Enlightenment hit upon an interesting question: Why is the texture of a leaf different on the top than on the bottom? Charles Bonnet (1720–1793), a Swiss naturalist, grew so obsessed with proving that the bottoms of leaves absorb and pump dew that he devoted an entire book to the topic. In a roundabout way, his quest led him to important early advances in phyllotaxis, including his descriptions of the quincunx and "redoubled spirals"—very likely the first mention of spirals in leaf arrangements. And although Bonnet was wrong in concluding that leaves function mainly to absorb dew, his successful detection of leaves' respiratory role spurred other scientists to the discovery of photosynthesis.

Bonnet came from a patrician French family that had moved to Geneva in the sixteenth century, fleeing anti-Protestant persecution. (At that time, Geneva was a Calvinist stronghold.) From early childhood on, Bonnet was partially deaf.

FIG. 4.1 Portrait of Charles Bonnet.

When his classmates were cruel, his parents hired private tutors to educate him at home. Under their encouragement, he developed keen powers of observation and was insatiably curious. At university in Geneva, he met Jean-Louis Calandrini, the Swiss mathematician and philosopher who would inspire Bonnet's work on leaves. Bonnet also corresponded with leading scientists of the time, including René-Antoine Ferchault de Réaumur, an engineer and entomologist who was known for inventing a temperature scale and once tried to make gloves from spider silk.

Initially, Bonnet studied insects—a pursuit that brought him much success. Advised by Réaumur, Bonnet discovered when only 20 years old that female aphids can reproduce asexually, from unfertilized eggs. This act of self-cloning is now called parthenogenesis. But by his early 30s, Bonnet had peered at so many caterpillars, tapeworms, and aphids through his microscope that his eyesight began to fail. "Deprived of what had until now been my greatest delight, I sought consolation in changing pursuits," he wrote in his book on botany. "I turned to Plant Physics, a subject less alive, less fertile in discoveries, but generally considered more useful."[1] Incidentally, Bonnet's impaired vision would also lead to the hallucination syndrome that bears his name (see below).

It was dew that led Bonnet to plant spirals. Dew was the central preoccupation of his 1754 book, *Recherches sur l'usage des feuilles dans les plantes* (Research on the function of leaves in plants). To find out whether leaves could absorb liquid, Bonnet devised experiments using simple components: glass test tubes, lead lids with holes, sunlight, water, olive oil, and powder boxes. If some of his experiments seem quaint to our eyes, it is important to recall that the scientific method itself was still in its infancy. More importantly, Bonnet was endlessly energetic in his pursuit of truth. He also wanted everyone to get outdoors and look closely at plants, writing:

> Not even the most perfect drawings can match Nature: she's the one you must consult. I wish to inspire that desire in my Readers, & to encourage them to go out into the countryside to find the originals of my mere copies. In this way, they will verify my Observations while wandering.[2]

## Do Leaves Absorb Dew?

Before we describe Bonnet's experiments in more detail, it's useful to know that Bonnet believed that dew comes *up* from the ground overnight. He stated very clearly that "dew rises from the Earth at the setting of the Sun."[3] This helps us understand why he thought the underside of leaves might have the function of pumping dew. It is also possible that he confused dew, which condenses on the tops of leaves, with fog or mist, which might cling to their bottom sides.

To test his hypothesis, Bonnet cut off leaves from various plants, smeared them with oil on one side, and placed them either face up or face down on the surface of water (see fig. 4.2). He then compared the number of days before the leaves "turned bad."

Bonnet ran his experiment on 14 species of "herbs"—including nettles and sunflowers—and 14 species of trees and bushes. For the "herb" species, results were mixed, as several lived longer when lying face down on the water. But the results for tree leaves were conclusive: for all but two species, the leaves lasted longer when face up. In some cases, the face-up leaves lasted weeks or even months.[4]

Based on his results, Bonnet felt confident in his conclusion: "Trees pump dew using the underside of their leaves."[5] He had seen that tree leaves last longer when the underside is placed on water, and to him this showed that leaf bottoms must absorb water (and therefore dew).[6] He concluded that

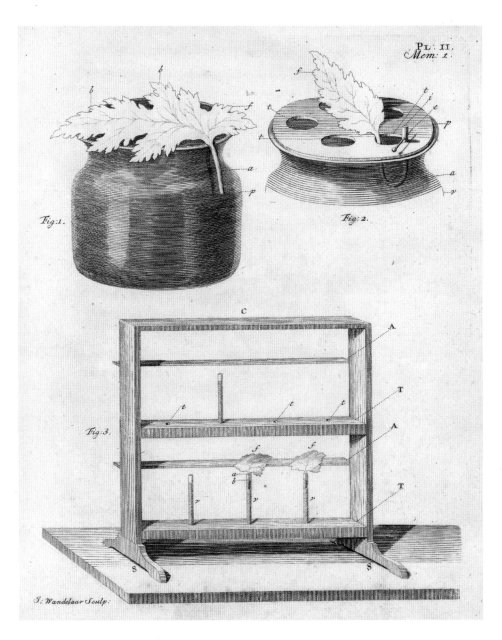

**FIG. 4.2** Illustrations of Bonnet's experiments on leaves and absorption. At bottom, two leaves of a four o'clock flower (*Mirabilis jalapa*) sit in glass test tubes. One stem is immersed in water and the other in olive oil. Which will pump more from the tube?

- 🌀 "The top side [of the leaf] mainly serves as defense or shelter for the bottom side, as it must face the sky or the air."
- 🌀 "The bottom side, having as one of its main functions the pumping of dew, must face the ground or the interior of the plant."[7]

Bonnet's experiments on dew pumping left some later scientists skeptical. (More recently, scientists have established that some plants do absorb dew through their leaves, mainly species that live in the desert.) But if his ideas on leaf undersides were shaky,[8] certain of his experiments turned out to be groundbreaking.

## Do Plants Breathe?

Bonnet wondered whether plants might breathe, and he devised a series of "experiments to discover whether leaves are the lungs of plants."[9] In the summer of 1747, he placed grape leaves, still attached to their vine, in empty *poudriers* (containers for powder, likely used for wigs). He filled the vessels with water and began making observations. "As soon as the sun began to warm the water in the vessels, I saw appearing on the leaves of the stems many bubbles, like little pearls."[10] He noted that the bubbles did not form at night.

Bonnet concluded: "One can infer that leaves do not only serve to pump water, but also to introduce into vegetable bodies much fresh and flexible air."[11] He went on to ask: "But do plants breathe?" None other than Bonnet himself had discovered the breathing pores of caterpillars and butterflies, the organs now known as spiracles. He wondered whether plants might also be covered with similar "stigmata," or have them only in certain places.

Bonnet's early experiments on plant respiration inspired scientists on the path to discovering photosynthesis, including Joseph Priestley, Alphonse de Candolle, and Antoine Lavoisier. A generation after Bonnet, the Dutch physician Jan Ingenhousz would describe the process in 1779, generally receiving credit for our modern understanding that the green parts of plants absorb carbon dioxide and release oxygen when in sunlight. Note too that Bonnet's leaf-under-water experiment has been re-created by countless schoolchildren in science class.

### The Five Leaf Orders

Having introduced Bonnet's experimental methods, we now move on to his discoveries in phyllotaxis. He found immense joy in the five orders he observed, writing, "I admit that I was astonished by these leaf arrangements."[12]

His chapter on phyllotaxis opened by underscoring what he had accomplished in his book thus far: "After everything that I just laid out about leaves, I am confident that not a single reader doubts that among their principal functions is the pumping of dew."[13] There remained, however, an additional problem to consider. If leaves overlap each other, then certainly that arrangement "interferes with the dew reaching the ones that are above."[14] At the same time, he wrote, transpiration in plants demands that "air circulate freely around" the leaves. They should cover each other as little as possible.

The solution given by nature, he observed, is that in many plants, leaves are arranged in spirals. Bonnet called it "one of those facts that is always right under our noses"[15] yet had somehow escaped attention.

Bonnet then laid out the five orders of leaf arrangement. The first three—*alternate*, *decussate* ("at crossed pairs"), and *whorled*—were already known.

In the fourth order, *quincunx*, the leaves come in groups of five, each leaf making 2/5 of a turn as it steps up from the previous one. This means that the sixth leaf appears exactly over the first one. Although the quincunx was not new (Leonardo da Vinci had already mentioned it), Bonnet described it more accurately.

For the fifth order, Bonnet attributed to his mentor Calandrini the discovery of *redoubled spirals*. To our knowledge, this is the first time the word "spiral" appears in connection to phyllotaxis. The leaves are evenly arranged in three groups of seven, along three parallel spirals or helices. On each helix, a leaf makes 1/7 of a turn as it steps up from the previous one.[16]

Having laid out this neat organization, Bonnet noted that he felt immensely satisfied. From that point on, he wrote, "I carefully observed any plant that fell into my hands, determining which order it fell into."[17]

In his book, he classifies 125 plant species according to his five orders, with the first 75 listed here. For each order, he separates the "herbaceous" plants from the "ligneous" (tree and shrub) ones.

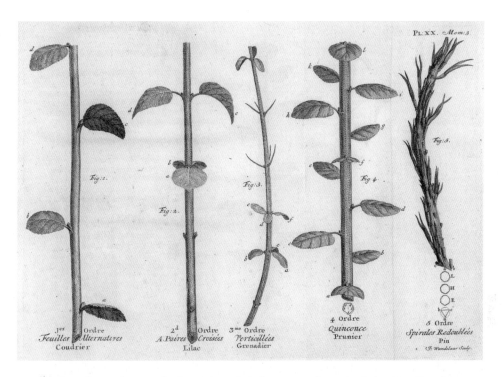

**FIG. 4.3** At top, Bonnet lays out the five orders of phyllotaxis, as drawn by Calandrini. The plants are labeled from left to right: hazel, lilac, pomegranate, plum, and pine.

At bottom, more examples of Bonnet's five orders as seen in the plant world. From left to right: cherry laurel (*Prunus laurocerasus*), waxleaf privet (*Ligustrum japonicum*), lemon verbena (*Aloysia citriodora*), white correa (*Correa alba*), and Norway spruce (*Picea abies*).

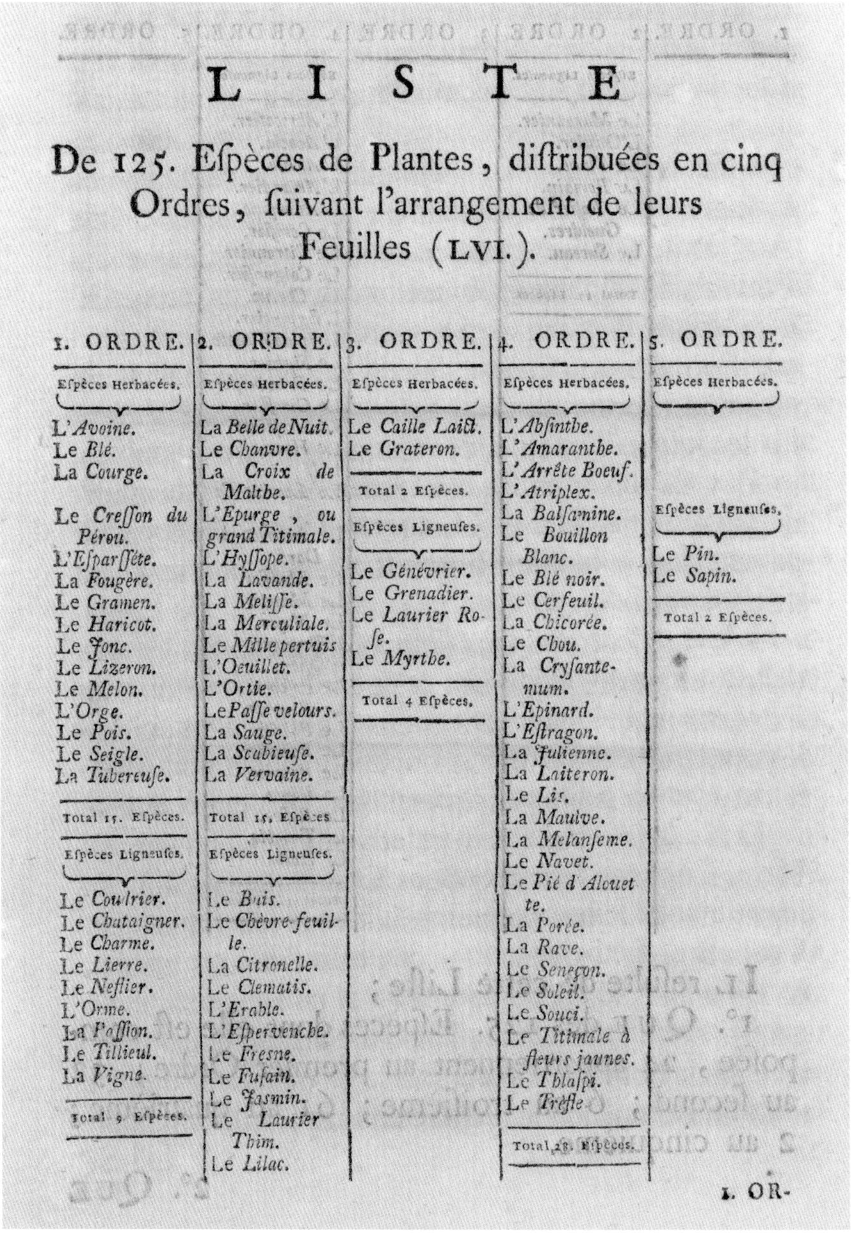

# LISTE

De 125. Espèces de Plantes, distribuées en cinq Ordres, suivant l'arrangement de leurs Feuilles (LVI.).

| 1. ORDRE. | 2. ORDRE. | 3. ORDRE. | 4. ORDRE. | 5. ORDRE. |
|---|---|---|---|---|
| Espèces Herbacées. | Espèces Herbacées. | Espèces Herbacées. | Espèces Herbacées. | Espèces Herbacées. |
| L'*Avoine.* | La *Belle de Nuit.* | Le *Caille Lait.* | L'*Absinthe.* | |
| Le *Blé.* | Le *Chanvre.* | Le *Grateron.* | L'*Amaranthe.* | |
| La *Courge.* | La *Croix de Malthe.* | Total 2 Espèces. | L'*Arrête Boeuf.* | |
| Le *Cresson du Pérou.* | L'*Epurge, ou grand Titimale.* | Espèces Ligneuses. | L'*Atriplex.* | |
| L'*Esparssète.* | L'*Hyssope.* | Le *Génévrier.* | La *Balsamine.* | Espèces Ligneuses. |
| La *Fougère.* | La *Lavande.* | Le *Grenadier.* | Le *Bouillon Blanc.* | Le *Pin.* |
| Le *Gramen.* | La *Melisse.* | Le *Laurier Rose.* | Le *Blé noir.* | Le *Sapin.* |
| Le *Haricot.* | La *Merculiale.* | Le *Myrthe.* | Le *Cerfeuil.* | Total 2 Espèces. |
| Le *Jonc.* | Le *Mille pertuis.* | Total 4 Espèces. | La *Chicorée.* | |
| Le *Lizeron.* | L'*Oeuillet.* | | Le *Chou.* | |
| Le *Melon.* | L'*Ortie.* | | La *Crysantemum.* | |
| L'*Orge.* | Le *Passe velours.* | | L'*Epinard.* | |
| Le *Pois.* | La *Sauge.* | | L'*Estragon.* | |
| Le *Seigle.* | La *Scabieuse.* | | La *Julienne.* | |
| La *Tubereuse.* | La *Vervaine.* | | La *Laiteron.* | |
| | | | Le *Lis.* | |
| | | | La *Maulve.* | |
| Total 15 Espèces. | Total 15 Espèces. | | La *Melanseme.* | |
| Espèces Ligneuses. | Espèces Ligneuses. | | Le *Navet.* | |
| Le *Coudrier.* | Le *Buis.* | | Le *Pié d'Alouete.* | |
| Le *Chataigner.* | Le *Chèvre-feuille.* | | La *Porée.* | |
| Le *Charme.* | La *Citronelle.* | | La *Rave.* | |
| Le *Lierre.* | Le *Clematis.* | | Le *Seneçon.* | |
| Le *Neslier.* | L'*Erable.* | | Le *Soleil.* | |
| L'*Orme.* | L'*Espervenche.* | | Le *Souci.* | |
| La *Passion.* | Le *Fresne.* | | Le *Titimale à fleurs jaunes.* | |
| Le *Tillieul.* | Le *Fusain.* | | Le *Thlaspi.* | |
| La *Vigne.* | Le *Jasmin.* | | Le *Trèfle.* | |
| Total 9 Espèces. | Le *Laurier Thim.* | | Total 28 Espèces. | |
| | Le *Lilac.* | | | |

I. OR-

**FIG. 4.4** Bonnet classified 125 plants according to five phyllotaxis orders. "I don't approach the subject as a Botanist, [but] as a simple Observer," he wrote. And in fact, he uses only the common French names for plants, not the scientific ones. Do you agree with his classification of the myrtle, "Le *Myrthe*"?

## Quincunx

A *quincunx* (QUIN-cunks) can refer to various patterns, including the arrangement of five dots on dice. But in phyllotaxis, the term has a very specific meaning: a spiral pattern with every sixth leaf aligned directly above the first one. Here is Bonnet's description:

> The fourth order can be called *quincunx* & is composed of leaves distributed five by five. To clearly conceptualize this distribution, draw on a stick five [vertical] lines that are parallel & at equal distance from each other. At the bottom of the first line, mark the place of the first leaf. A bit above, & on the third line, place the second leaf. At equal [vertical] distance from this one, & on the fifth line, put the third leaf. Place the fourth leaf on the second line: the fifth leaf will occupy the fourth line. We will thus have a sequence of five leaves that do not overlap.[18]

On the diagrams in figure 4.5, the black and gray points are the leaves, drawn as Bonnet suggests on five vertical lines, skipping one vertical line at each step. The sixth leaf returns to the starting line, following two full turns around the cylinder.

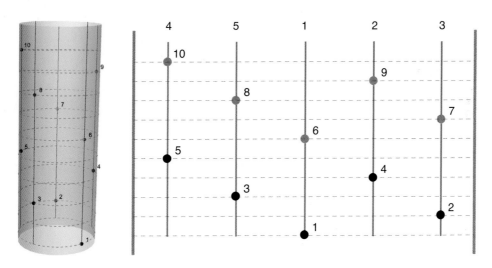

**FIG. 4.5** Construction of the quincunx. At left, the quincunx structure is shown on a cylinder—like the stick that Bonnet describes. At right, the same structure is shown "unrolled" on a plane. The thick gray lines at left and right represent the edges where the cylinder meets.

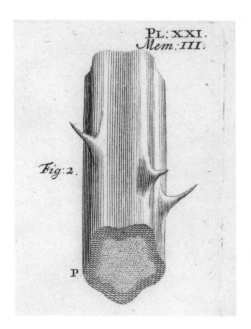

**FIG. 4.6** Note how in figure 4.3, Bonnet has put a tiny pentagon below the plum branch illustrating the quincunx order, showing five vertical lines. He observed that these lines sometimes correspond to ridges on a stem, as seen at right in the neat pentagon of the bramble plant. Is it a regular pentagon?

Bonnet continues up the stick, placing a similar group of five leaves above the preceding group: leaf 6 sits directly above leaf 1. In this way, a leaf appears every 2/5 of a turn around the helix. Bonnet does not talk about a helix here, however—or even a spiral. But he's getting there.

### Redoubled Spiral

Bonnet credits his mentor Calandrini with "great wisdom" for discovering what they called "redoubled spiral patterns." Bonnet gives directions for constructing a redoubled spiral:

> Trace on a stick three or five parallel spirals; on each turn of these spirals, pin at an approximately equal distance from one another seven or eleven pins, & you will have a very precise idea of this arrangement.[19]

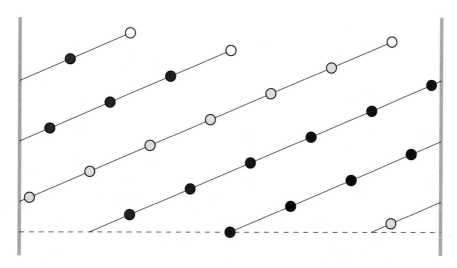

**FIG. 4.7** A redoubled spiral, as Bonnet described it. Here the plant stem is again shown as an "unrolled" cylinder with a point for each leaf. The three spirals—red, blue, and yellow—are connected by light gray lines. How many other spirals link the different-colored leaves?

To trace the redoubled spiral in the image above, start with the blue spiral. Each blue point makes 1/7 of a turn around the cylinder from the previous point. When the spiral crosses the right edge of the cylinder, it continues going up on the left side. The eighth point, in white, sits directly above the first (lowest) blue point. The same rules apply to the yellow and red spirals. As Bonnet noted, after 21 points (3 × 7), the pattern repeats.[20] He also described a similar pattern based on 5 spirals with 11 points per turn (making 5 × 11 = 55).

In his observations, Bonnet noted the groups of numbers that occur in both types of redoubled spirals. "In the first case, one counts 21 leaves in a complete turn of the 3 spirals," he wrote, and "in the second case, the complete turn of the 5 spirals gives 55 leaves."[21] All of these are Fibonacci numbers, which is no mere coincidence, as we will see later.

To his credit, Bonnet did not shy away from describing exceptions to his five classifications. (For some intriguing details, see the appendix.) These details not only reveal his honesty as an observer but also make his work more universal and ahead of its time. He noted that, in plants following the quincunx pattern, the sixth leaf is often not quite aligned with the first—and that therefore the more vertical rows actually spiral uniformly around the stem, rather than running straight along it.[22] Bonnet also found that some plants displayed more than one order, at different places along the stem. And as for corn, it did not fit at all into any of his five orders. The corn mystery would not be solved until the twenty-first century (see chapter 14).

And so in the eighteenth century, the study of phyllotaxis took a jump forward, as Bonnet and his mentor Calandrini became the first to write explicitly about plant spirals. While Fibonacci numbers loom large in their observations, they never spoke of the numbers as a sequence. Not until the following century would Schimper, Braun, and the Bravais brothers revisit these spiral structures and uncover the Fibonacci numbers within.

**Coda: Bonnet Syndrome**

Bonnet described the visual hallucinations following vision loss—now known as Charles Bonnet syndrome—in 1760, based on the case of his elderly grandfather, who had undergone cataract surgery. Bonnet wrote:

> I knew a respectable man, in good health, with sound judgment and memory, who, in plain daylight and without any outside influence, sometimes saw before him figures of men, women, birds, carriages . . . [and] saw buildings rise before his eyes, with all their architectural details. The tapestries in his apartment seemed to transform into different and more elegant ones.[23]

Later, when Bonnet's eyesight deteriorated, he too experienced these hallucinations. He described the condition as "how the theater of the mind could be generated by the machinery of the brain."[24]

Three centuries later, neurologist Oliver Sacks opened his book *Hallucinations* with Rosalie, a woman in her 90s who had become blind. She described the silent figures she "saw" walking back and forth, wearing exotic clothing, and taking no notice of her. Was she losing her mind? Dr. Sacks assured her that she was not—and told her about Bonnet.

# Try Your Hand

## Make a Spiral Stem

This activity provides hands-on practice in building a simple Fibonacci "stem" with 11 spiraling leaves, using a cardboard toilet paper tube. You'll also need a 20″ length of string, tape, a marker, and a sheet of paper or thin cardboard (green if possible) for the leaves.

a. Draw two dots directly above one another on the tube, one near the top and one near the bottom.
b. Tape one end of your string to the bottom dot.
c. Starting at the bottom, wind the string evenly and snugly around the tube, making four full turns and ending at the top dot. Cut the string where it meets the top dot.
d. Remove the string from the tube. Measure it (centimeters might be easier here than inches). Divide the result by 11. This is your leaf interval.
e. Using this interval, measure and mark 11 dots along the string—one at each end and nine in between. The dots indicate where the leaves will be attached.
f. Now, once again tape one end of the string to the bottom dot. Wind the string evenly back up to the top, again making four full turns. This time, tape down the other end of the string to the top dot.
g. Cut out 11 paper leaves, each about 2″ long. Tape them to the dots. Bend them outward to make them look more leaflike.
h. What is the angle between two consecutive leaves?
i. Is this model a good fit for Bonnet's quincunx, illustrated in figure 4.8?

**FIG. 4.8** Does your cardboard stem fit this model?

Extra credit: Make another stem the same way, but this time winding six full turns around the stem instead of four. Does this one fit Bonnet's model? What about Karl Schimper's model in the next chapter?

**Draw Bonnet's Redoubled Spirals**

In his book on leaves, Bonnet describes a "redoubled spiral" with three spirals winding up the plant stem. In figure 4.7 above, we illustrate his concept using three sets of colored dots joined by lines. But Bonnet also describes a similar pattern based on *five* spirals. In this activity, each of the five colored lines represents a spiral (helix) going up a cylindrical stem.

1. Photocopy the chart in color. Start with the blue line. Draw a point every five rungs of the ladder, starting from the bottom rung (where the blue line intersects the bottom gray line). Your points should be at rungs 1, 6, 11, etc. After the sixth point, when you reach the right edge of the paper, jump to where the blue line restarts on the left edge and keep going. The 12th point you draw should be the last one that fits, and it should be directly above your first point.
2. Now do the same on the lower purple line (skipping over the red line), this time placing points every five rungs but starting from the *second* rung.
3. Repeat the process using the orange, red, and green lines, in that order, each time starting on the next higher rung. For the orange line, start on rung 3. For the red line, start on rung 4. And for the green line, start on rung 5.
4. Check that you have 12 points for each spiral, with the last point directly above the first. For each of the five spirals, there should be 11 points per turn, for a total of 55. (Right, a Fibonacci number!)
5. Roll your completed chart into a cylinder, matching up the colored lines from one edge to the other. Tape the edges together. And voilà! This is the 3D model "stem" that Bonnet offered his readers as a thought experiment.

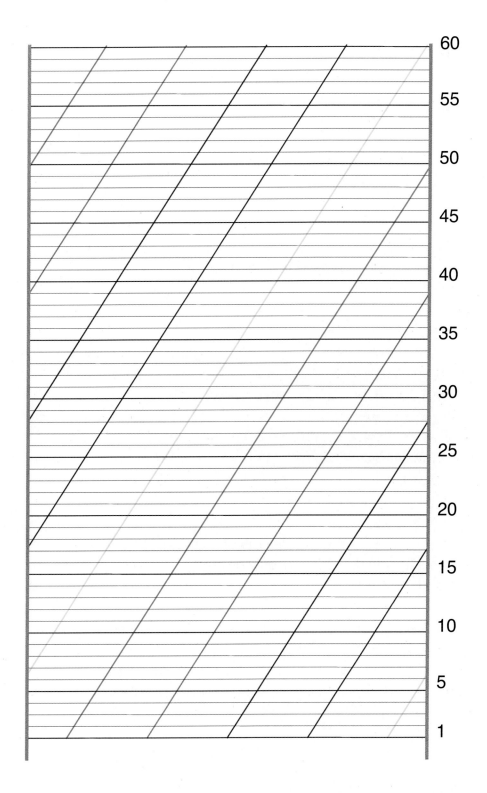

**FIG. 4.9** Chart for making five redoubled spirals.

## CHAPTER 5

# Biomathematics on a Watch Face

Math
met
poet!
Schimper swooned
o'er Fibonacci.
He coined the word phyllotaxis,
it seems—but failed to publish and people stole his fire.

Theirs was a fruitful friendship . . . until everything unraveled. At Heidelberg University, Karl Schimper (1803–1867) and Alexander Braun (1805–1877) were inseparable.[1] Both brilliant students, they had grand ambitions. Together they developed a way to analyze plant geometry using continued fractions, the first known application of math to the study of plants. As Julius von Sachs wrote in his history of botany, their findings "could be expressed in numbers and formulae—a thing hitherto unknown in botanical science."[2] Botanical biomath had begun.

Also under the gazes of Schimper and Braun, for the first time Fibonacci numbers would be placed at the heart of plant patterns. Both scientists would ask the obvious question: Why do these numbers appear? Based on their early observations, the two frenemies would offer somewhat different interpretations.

Schimper was not only a gifted scientist but also a romantic poet. His wordsmithing left a lasting mark on plant mathematics, for, we believe, in 1830 he coined the word "phyllotaxis."[3] Based on the Ancient Greek words for "leaf arrangement," the term quickly caught on. With the name "phyllotaxis," the new field claimed its own specialized vocabulary, one that transcended national borders. The old German and French terms, *Blattstellung* and *arrangement des feuilles*, would soon disappear.

And that was not all: Schimper also coined the important term *divergence angle*.[4] Together, he and Braun did groundbreaking work describing the concept in plants, as we will see in a moment. As an aside—but a memorable one—in 1837 Schimper came up with the concept of the Ice Age and coined that term as well.[5] In his usual self-defeating way, he announced his discovery not in a scientific paper, but in a poem.[6] Other scientists took his idea and got the fame.

FIG. 5.1 Portrait of Karl Schimper.

For all his talent, Schimper as a scientist never went very far, mainly because he rarely got around to publishing what he had discovered. Born in the German city of Mannheim, he was the child of a math teacher father and a baroness mother. When his mother left and the couple divorced, Schimper was left to grow up in poverty. He was able to attend university only by studying theology, the sole discipline that offered a stipend.

At Heidelberg University in the 1820s, Charles Bonnet's plant spirals suddenly became a hot topic, after being neglected for decades. Schimper and Braun got fired up by his ideas, and they made impressive progress describing the complex geometries they saw in plants. Schimper was the first to make their discoveries public—an announcement that he knew would make a splash. But in typical fashion, Schimper hid his phyllotaxis breakthroughs. He nested them inside a botany paper that appeared to be completely unremarkable, discussing how to distinguish various species of comfrey.

### Introducing Divergence Angles on a Watch Face

When Schimper finally got to the heart of his groundbreaking paper, he introduced the topic by way of his comfrey plants:

> Now I want to show that the leaves of these plants have a *definite* regularity, a regularity that even has a name . . . , and that can be described mathematically, with a numerical expression—and not only in these plants but in the entire plant world.[7]

Although Schimper did not credit Bonnet, he now expanded on Bonnet's geometric construction of the quincunx. Like Bonnet, he drew equally spaced vertical lines on a cylindrical stem. Then he placed leaves as points at regularly increasing heights. Again like Bonnet, Schimper constructed the spiral pattern by placing the leaves one by one, passing a given number of lines before placing a new one.[8]

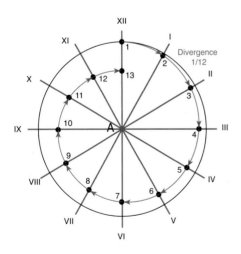

**FIG. 5.2** Schimper introduced his divergence angles with 12 "direction-lines" around a stem, using a simple watch face system. (This is a modern recreation, as Schimper did not publish illustrations.)

There was a minor problem, however. Schimper's paper, perhaps hastily sent off to the press, lacked any illustrations to demonstrate his points. And so he turned to something his readers would have on hand: their pocket watches. To explain his plant spirals, he used a watch face model to represent a stem, divided into 12 sections by what he called "direction-lines."[9] Along the stem, one leaf would grow on each direction-line.

More importantly, Schimper used his watch face model to illustrate the steps *between* leaves,

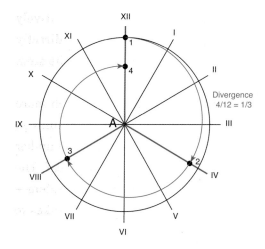
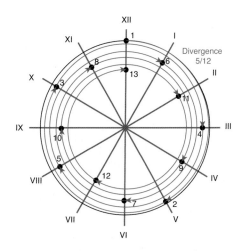

**FIG. 5.3** Here, Schimper took four steps before reaching the next leaf. How many lines are covered by leaves?

**FIG. 5.4** This time, Schimper crossed five direction-lines between each point on the spiral, making the divergence angle = 5/12.

in order to find the divergence angle. Drawing a spiral, Schimper would count the number of direction-lines crossed from one point (or leaf) to the next. In the first case he discussed, figure 5.2, he simply took one step on the watch face, so the number of direction-lines between points was 1. This made the divergence angle = 1/12.

In his next example, figure 5.3, Schimper passed 4 direction-lines between points (or leaves), making the divergence angle = 4/12 or 1/3. (Note that there are only 3 purple direction-lines with leaves.[10] This is because the fourth leaf grows directly above the first.)

The final case that Schimper discussed was the most interesting. This time, he crossed 5 direction-lines between each point on the spiral, marking the first point (or leaf) at 12 o'clock, the second at 5 o'clock, the third at 10 o'clock, the fourth at 3 o'clock, and so on. The divergence angle = 5/12. Since 5 and 12 are relatively prime, he made stops at all 12 direction-lines before returning to where he started, having made 5 full turns around the stem. ("Relatively prime" means they lack a common divisor. Here, the 13th leaf grows directly above the first, just as in fig. 5.2 above—but the spirals winding up the stem are quite different.)

With this system, Schimper took Bonnet's original idea to a much more powerful place by generalizing it. Bonnet had drawn only five vertical lines on the stem, placing leaves on every other line. Schimper draws a general number of $n$ vertical lines on the stem, with $p$ lines passed to reach the next leaf. The fraction $p/n$ defines the divergence angle, in units of turns around the stem—with $n$ the number of leaves in a cycle, and $p$ the number of turns it takes to pass through them.

## Generative Spiral

Here we make a quick stop to mention another new term introduced by Schimper. As we have seen, he constructed his spiral patterns systematically, placing successive leaves along the stem according to a constant divergence angle. In the 1830 edition of his phyllotaxis article, he named this spiral the *Wendel*, taking the term from the German word *Wendeltreppe*, or spiral staircase.[11] Ever the poet! (Technically speaking, as Chris the mathematician will tell you, the staircase is not a spiral but a helix. In this book, however, as stated in the introduction, we will follow convention and generally use the term "spiral.") At any rate, in English Schimper's construction is now called the *generative spiral*—the spiral that generates all the visible spirals (parastichies). We will need this in the Braun chapter, coming up next.

## Enter Fibonacci

After measuring the divergence angles in hundreds of real plants, Schimper discovered that most of these angles fall within the following sequence:

$$\frac{1}{2}\text{(alternate phyllotaxis)}, \frac{2}{5}\text{(quincunx)}, \frac{3}{8}, \frac{5}{13}, \frac{8}{21}\text{(redoubled spiral)}\ldots$$

Notice that all are quotients of Fibonacci numbers.[12] While he did not indicate that he had any prior knowledge of the Fibonacci sequence—or its relation to the golden ratio—Schimper rediscovered some of the sequence's properties while attempting to explain this phenomenon. His reasoning, even if incomplete, may well be the first time a scientist combined math and biology. And he launched a centuries-long chain of attempts to explain Fibonacci phyllotaxis that continues to this day.

Although Schimper's 1830 paper has no figures, it does include a series of tables displaying all the mathematically possible divergence angles. Schimper then tries to explain *why* Fibonacci quotients appear so often in nature by introducing a somewhat strange average of fractions, obtained by adding together numerators and denominators.

In laying out his theory, Schimper starts by stating that the largest divergence angle found in plants is $\frac{1}{2}$. (A larger divergence, say $\frac{2}{3}$, gives the exact same pattern as a $\frac{1}{3}$ divergence, which is $= 1 - \frac{2}{3}$.) He also notes that he has not seen any plants with divergence angles of less than $\frac{1}{3}$, as the leaves would be too close together. So the only two possible divergence angles that can appear within one full turn of the stem are $\frac{1}{2}$ and $\frac{1}{3}$, which he places in the first row.

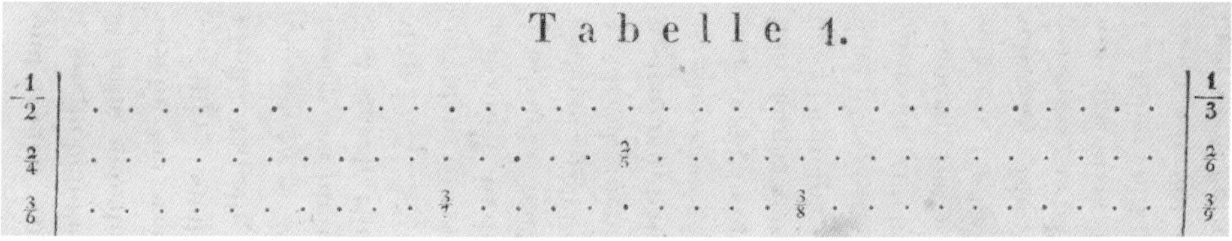

**FIG. 5.5** The first rows of Schimper's table of divergence angles.

Next, he finds all possible divergence angles that can appear within two full turns of the stem, in the same range of $\frac{1}{2}$ to $\frac{1}{3}$. These are $\frac{2}{4}, \frac{2}{5}$, and $\frac{2}{6}$. The first and last, equal to $\frac{1}{2}$ and $\frac{1}{3}$, are already accounted for. So $\frac{2}{5}$ is the only new

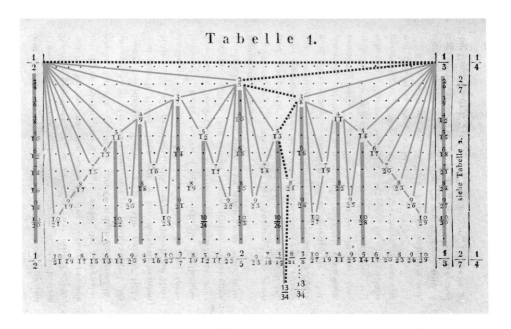

**FIG. 5.6** Schimper's table of divergence angles, organized according to increasing numerators. Annotations show, with red dashes, the Fibonacci progression of divergences. (Note that 13/34 was originally misplaced.) Do you see curves joining the divergences?

possibility. He notes that this is in fact a very common divergence angle in plants, without mentioning its name—our old friend the quincunx: cycles of five leaves in two full turns around the stem, returning to where we started.

Similarly, the possible divergence angles found within three turns of the stem in the chosen range are 3/6, 3/7, 3/8, and 3/9, with only the middle two being new. He then proceeds in the same fashion to complete a table of all possible divergence angles of up to 10 turns.

Mathematically, this tree of fractions was intriguing—but Schimper knew from his plant observations that not all of these divergence angles occur in nature. (For more on fraction trees, see the appendix.) He saw that the divergence angles in most real plants involve Fibonacci numbers, and he also saw a way to generate them. In order to obtain divergence angles with more leaves

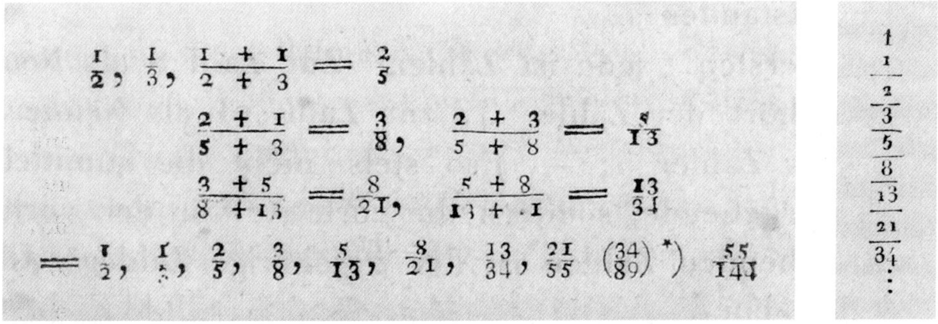

$$\frac{1}{2}, \frac{1}{3}, \frac{1+1}{2+3} = \frac{2}{5}$$

$$\frac{2+1}{5+3} = \frac{3}{8}, \quad \frac{2+3}{5+8} = \frac{5}{13}$$

$$\frac{3+5}{8+13} = \frac{8}{21}, \quad \frac{5+8}{13+21} = \frac{13}{34}$$

$$= \frac{1}{2}, \frac{1}{3}, \frac{2}{5}, \frac{3}{8}, \frac{5}{13}, \frac{8}{21}, \frac{13}{34}, \frac{21}{55}, \left(\frac{34}{89}\right)^*, \frac{55}{144}$$

$$\frac{1}{1}, \frac{2}{3}, \frac{3}{5}, \frac{5}{8}, \frac{8}{13}, \frac{13}{21}, \frac{21}{34} \ldots$$

**FIG. 5.7** At left, Schimper lists Fibonacci divergence angles, based on adding numerators and denominators to obtain a "mediant" average. At right, he shows the Fibonacci sequence.

and more turns, he added together the numerators and denominators of the previous two fractions to get the next one. This funny average is now known as the "mediant":

$$\frac{a}{b}, \frac{c}{d} \to \frac{a+c}{b+d}$$

Schimper takes this mediant as a real average between the two previous phyllotaxis orders: if the first one places $a$ leaves in $b$ turns, he puts above it the second one with $c$ leaves in $d$ turns. In total, this leads to $a+c$ leaves in $b+d$ turns. To Schimper, this represented a "new and unexpected law of Nature." Starting from 1/2 and 1/3, he built the whole sequence of Fibonacci divergence angles, shown in figure 5.7.

To explain why this special series of fractions occurs in nature, Schimper stated that the leaves found themselves in a kind of conundrum, simultaneously repelling and attracting each other. He wrote:

> Just as there is a tendency for the leaves to form independently, moving as far away from each other as possible and isolating themselves, yet at the same time, the necessity to produce the most leaves possible makes them

form close to one another. . . . The divergences . . . are the result of a balance always sought between those two tendencies.[13]

For him, these tendencies of attraction and repulsion were represented by the two extreme angles, 1/3 and 1/2. Accordingly, the sequence of Fibonacci divergences can be seen zigzagging between these two angles, moving down his table (see fig. 5.6). While Schimper notes that these fractions tend to get closer and closer in value, he does not insist on this fact. As we will soon see, the Bravais brothers believed this limiting value to be very important—for better or for worse.

Schimper had grand plans to turn his paper into a book that would lay out his whole theory of phyllotaxis in detail. But he never finished the project. Finally, in 1835 his original paper was reissued in slightly expanded form. But Schimper considered this mere reworking a failure.

Schimper did eventually publish a collection of poems in 1840, and perhaps that brought him a measure of satisfaction. He was engaged twice—the second time to one of Braun's sisters—but neither engagement led to marriage. Nor did he obtain a professorship in botany, despite his friends' best efforts. In the end, he eked out an existence outside Mannheim, supported from 1845 on by a modest annuity paid by a local grand duke.

These lines from Schimper's poem "Theme" reveal his delight in uncovering the secrets of plants:

> Flowers turn their faces to me,
> Each one knows something and confides;
> From flowers, light shines on what is new,
> And so I learn a thing or two.[14]

If Schimper failed to achieve his potential, his friend Alexander Braun was marvelously productive—a factor that contributed to the collapse of their collaboration. Braun would go on to publish a gorgeously illustrated book expanding on Schimper's observations, focusing mainly on one example of spirals in the plant world: pinecones.

## Try Your Hand

### Write a Fibonacci Poem

Since Schimper was as much poet as scientist, this chapter seems the right place for a poetic exercise. Perhaps in childhood, you tried your hand at composing haiku. As you may recall, the form of the poem is based on the number of syllables per line, with the classic three-line haiku being 5-7-5 and alluding to the season.

A Fibonacci poem is also based on counting syllables—but in this variation, each line fits the mathematical sequence. Instead of the neat haiku triangle, a Fibonacci poem forms a staircase against a wall. Or if the words are centered, the poem makes a kind of fountain.

In this book, many chapters open with a seven-line Fibonacci poem whose last line has 13 syllables. In 2006, poet Gregory K. Pincus created a popular variation called a "fib," with just six lines (20 syllables) altogether. Like all Fibonacci poems, fibs fall at the pleasing intersection of language and math.

> Why (1)
> not (1)
> write a (2)
> poem named (3)
> for Fibonacci? (5)
> It might spiral out of control . . . (8)

CHAPTER 6

# So Many Spirals on a Pinecone

What's
in
pinecones?
Braun observed
more than meets the eye.
With his book, he lost a friend but
found a place for his sister's illustrations, signed C.

While Schimper procrastinated, his friend Alexander Braun forged ahead with his phyllotaxis work. Braun, who grew up in Germany near the Black Forest, had developed an early love for natural history and published his first scientific paper at the age of 16. If he and Schimper began on the same phyllotaxis page, in the end Braun saw the link between spirals and Fibonacci numbers in a very different light.

After Schimper failed to expand on his 1829 paper, Braun came out with his thick and gorgeously illustrated book on phyllotaxis in pinecones just two years later. The book has a long German title that translates to *Comparative Study of the Order of Scales on Pinecones, as an Introduction to the General Study of Leaf Arrangement*.[1] (Note that he did not yet use the word "phyllotaxis," as it seems Schimper had only just coined it.)

Although Braun thanked Schimper for his contributions on the very first page of the book, that was not good enough for his friend. Schimper was outraged that Braun had not shown him the manuscript before it went to press, no doubt wanting to stake his claim to the "mediant" idea. Braun's apology fell flat. Believing he deserved full credit for the theory, Schimper broke off all contact with Braun (though the two would eventually reconcile many years later).[2] History has united them, however: their work is generally known as Schimper and Braun's spiral theory of phyllotaxis.

With the success of his book, Braun would go on to a brilliant career as a professor of botany in Berlin, where he was beloved by his students. He is perhaps best known for work on cell theory, clarifying the definition of the cell as cytoplasm enclosed by a flexible membrane.

**FIG. 6.1** Cecilie (Cécile) Braun illustrated her brother Alexander's pinecone book. This is her self-portrait from 1829.

## Cecilie Braun's Exquisite Pinecones

In addition to having a mathematical mind, Braun had a practical bent. He composed his table of fractions to show real values of the divergence angles—and then explained how to observe these fractions in actual pinecones, which is not entirely obvious. If you're working with a plant whose leaves are clearly spaced along a stem, it's easy to label each leaf's "birth order" and find the divergence angle. (As Schimper had already explained, you count the number of direction-lines passed on a spiral, until you return to where you started.).

But how do you count the tightly packed scales on a pinecone? Needing a fresh approach, Braun turned to a geometric feature of pinecones that had never been described before.

**FIG. 6.2** Cecilie Braun drew these finely detailed pinecones for her brother's 1831 book, ranking her among the early women scientific illustrators to see her works published. How many spirals do you see?

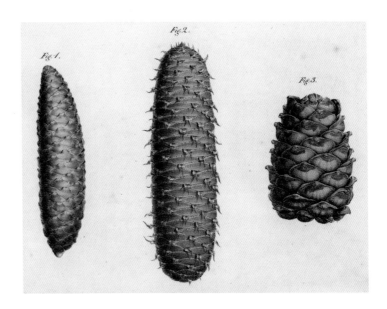

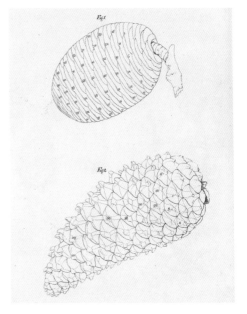

**FIG. 6.3** Here, Cecilie Braun illustrates cones from a cedar of Lebanon (*Cedrus libani*) at top, and a maritime pine (*Pinus pinaster*) below. At left, the cones appear as in nature, and at right with their scales numbered. How would you number the rest?

## The First Parastichy

To establish the order of pinecone scales, Braun starts by noting:

> The first thing we observe when looking at the surface of the whole cone is that the scales are arranged in oblique rows, which, as you can see if you follow them, describe helical lines around the cone.[3]

Each helix, which winds like a spiral staircase through adjacent scales, is now known as a parastichy. Using his sister Cecilie's beautiful and instructive sketches of actual pinecones, Braun showed how to draw the parastichies and count them.[4] (As we have already noted, following convention we will generally refer to these helices as spirals.)

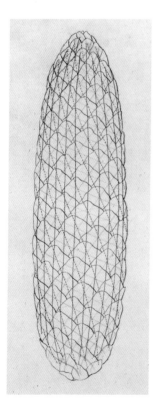 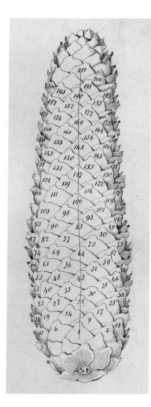

**FIG. 6.4** At left, Braun lays out his method for analyzing a pinecone (actually in this case a spruce cone). The cone's many spirals are drawn in three colors, most obviously the five red (left-handed) and eight blue (right-handed) ones. Also shown are three flatter spirals in yellow, 13 dashed more-vertical diagonals, and 21 solid vertical lines. Do these numbers look familiar? Could you guess from this picture alone that there are roughly two yellow spirals and five red ones?

At right, Braun shows his approach to numbering scales. The vertical spiral, shown here as the solid black line, passes through scales 1, 22, and 43, increasing each time by 21. How many vertical spirals are there?

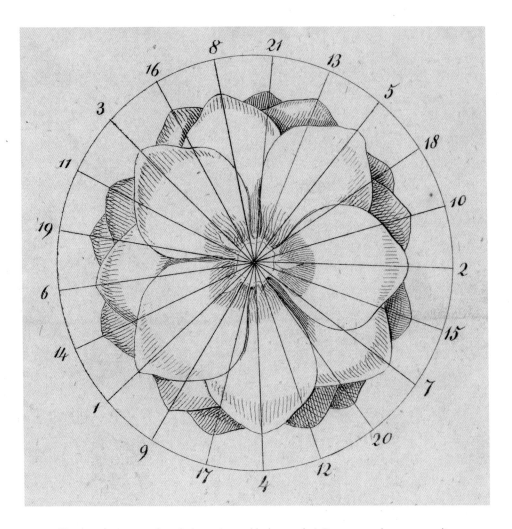

**FIG. 6.5** The same cone seen from below, trimmed below scale 1. Do you see the constant divergence angle between successive scales? Can you count the same number of rays between consecutive scales, just as Schimper proposed on his watch face?

Next, Braun shows how to determine the cone's divergence angle. His pinecone has 21 vertical spirals, some of them shown in figure 6.4 at left.[5] Also, looking at his cut cone in figure 6.5, you see 21 scale tips, indicating 21 directions. Braun assumed that this was a nice orderly pinecone specimen, whose scales turned regularly along the generative spiral and had a constant divergence angle. This allowed him to number the scales according to birth order, as shown in the rays. (The youngest scales are at the bottom of the pinecone. Scale 1 is located at about 7 o'clock, and scale 2 is at about 3 o'clock.) To get from scale 1 to scale 2, you pass through 8 vertical spirals. (More generally, this applies for any 2 consecutive scales, moving from scale $n$ to scale $n+1$.) Taking the 21 vertical spirals in total—before you return to the place you began—the divergence angle is 8/21.

## A Cool Trick for Numbering Scales

Braun also found a simple trick for linking the number of spirals to the individual scale numbers, when labeled according to birth order. Looking again at the cut pinecone in figure 6.5, Braun saw that starting from scale 1 (around 7 o'clock), the neighboring scale going clockwise is 6. Proceeding along the same spiral, next comes scale 11. There is a clear pattern here: $6-1=5$, and $11-6=5$. Continuing in the clockwise direction, you can see that in the adjacent spiral, starting from scale 3, the same pattern emerges: $8-3=5$.

In total, there are 5 similar clockwise spirals around the stem (starting with the first 5 scales, in the order 1, 3, 5, 2, 4). Braun recognized that these 5 spirals correspond to the 5 red spirals we saw in the left-hand pinecone of figure 6.4. He also observed that the difference in the birth order of scales along all these spirals is always 5.

This led Braun to propose a general rule: along a spiral, each scale's number *increases by the number of spirals in the chosen direction.*

We can easily apply this rule to the right-hand pinecone in figure 6.4 to see how Braun numbered all the scales. Start by finding scale 19, toward the bottom left. This cone has 8 blue right-handed spirals, so the number of each scale along this spiral increases by 8: 19, 27, 35, 43, . . . The same rule applies to the 5 red spirals going the other direction, whose scale numbers increase by 5: 25, 30, 35, 40, 45, . . . In this elegant way, counting the number of spirals leads directly to numbering the scales. (To try this out yourself, see the activity at the end of the chapter.)

## How Many Vertical Spirals?

In addition, Braun found a simple trick for determining the cone's total number of vertical spirals—now known as *orthostichies* (while we pronounce this OR-thoe-stick-ies, others say or-THA-stick-ies). To find this number, first decide which line of scales is closest to being vertical. Then subtract the number of a scale on this line from that of its lower neighbor.

We can try this out using figure 6.4, where Braun has drawn a line showing the most vertical spiral through scales 1, 22, 43, and 64. When we subtract, 64 − 43 = 21. Or in the same way, 43 − 22 = 21. The result means that there are 21 vertical spirals on this pinecone—just as Braun has drawn them, as the vertical lines in figure 6.4 and the rays in figure 6.5.

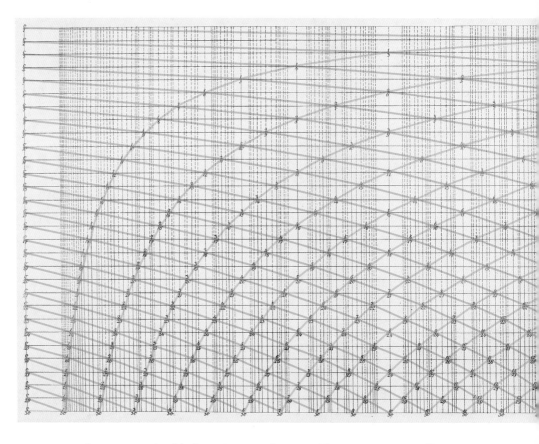

**FIG. 6.6** Braun's divergence angle table shows the precise location of the fractions, arranged by constant denominator. The red and green lines, added to emphasize the table's full symmetry, look a bit like parastichies.

Braun found other clever tricks hidden within the pinecone as well, related to the flattest (generative) spiral and divergence angles.[6] (For more details, see the appendix.) But it was his tree of fractions that really made his friend Schimper mad.

**Braun's Fraction Tree**

Like Schimper, Braun also created a tree of fractions showing every mathematically possible divergence angle within a given range.

In figure 6.6, Braun's fraction tree starts with 0/1 at top left, and 1/1 at top right. In contrast to Schimper's table, whose range was only between 1/2 and 1/3,

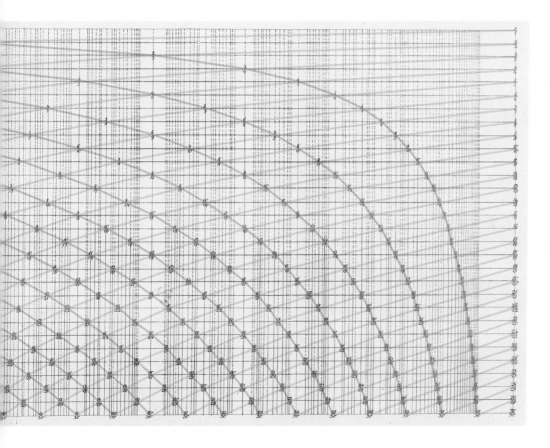

Braun's table between 0 and 1 reveals the full symmetry. Note, however, that Braun's table has somewhat less relevance to plants, and more to math.

In its vertical ordering, Braun's table is symmetric to Schimper's—the difference being that Schimper organized the fractions by increasing denominators as you go down the table, while Braun organized according to increasing numerators. Ultimately, Braun's approach would become the standard one. Sadly, though, mathematicians never discovered his work on fraction trees, which was known only to botanists.

It is important to note that Braun's version provides more information than Schimper's. Schimper shows only the link between a pair of fractions (with the mediant situated between the two). But Braun draws a plot of points, showing the exact position of each fraction along the interval between 0 and 1. In this way, he reveals precisely how the distance between the fractions changes each time you take a mediant, leading to these emptier or denser parts of the table in figure 6.6.

### Braun's Spirals in Radial Form

In the eye-catching radial view shown in figure 6.7, the pinecone's newer scales appear close to the center.[7] Older scales appear to move away from the center at constant speed. (In this case, they form Archimedean spirals.[8]) Note that scale 1, which is the oldest, appears at 6 o'clock. The youngest scale, scale 50, sits very close to the center of the circle. Counting the spirals from their starting points near the center, there are 3 yellow, 5 red, 8 blue, and 13 thin black. In addition, 21 straight black lines mark the vertical orthostichies, just as we would expect. Braun has also drawn in the diagonal relationships that connect the spirals.

In figure 6.8's highly original view, Alexander Braun sliced the bud of a radish (*Raphanus sativus*) through the middle, revealing its 3/8 phyllotaxis. To calculate the divergence angle, start at 1, which bisects the oldest and largest leaf. Then rotate 3 steps counterclockwise to find 2. Continue this process to reach the line that bisects leaf 8. The next leaf in the series, which would be labeled 9—the second smallest leaf—is almost (but not quite) aligned with 1.[9] Note that at the fifth leaf, the leaves start to stray from their prescribed alignments.

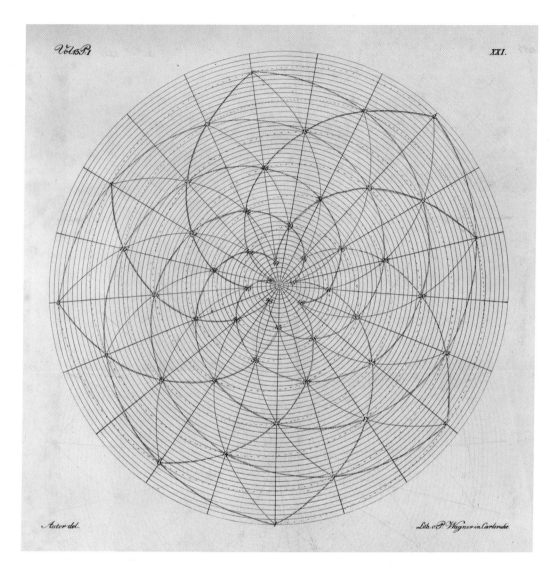

**FIG. 6.7** Here, Braun draws a schematic representation of the same pinecone scales shown in figure 6.5, but this time in radial view. The parastichy numbers are (5, 8).

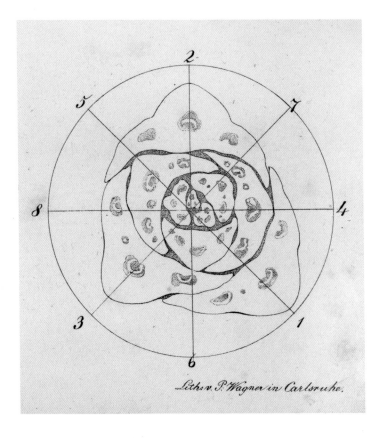

**FIG. 6.8** Very likely, this drawing is a scientific first. Cecilie Braun has illustrated her brother's discovery of phyllotaxis in a bud. Do you agree with the numbering of the young leaves?

When Schimper and Braun first presented their breakthroughs starting in 1829, "their fellow botanists applauded the originality and elegance of the spiral theory,"[10] as William Montgomery noted in his phyllotaxis history. But future researchers would ultimately reject Schimper and Braun's concept of rigidly fixed geometries. Moreover, neither Schimper nor Braun ever wrote proofs for their more mathematical statements. (This is especially sad for Braun, whose very original mathematical observations got buried in his long pinecone book.) For these proofs, we will need the wide-ranging mind of the scientist featured in the next chapter, adventurer and crystallographer Auguste Bravais. To explain Fibonacci phyllotaxis, Bravais and his brother would use the spirals seen by Braun in a very different way, bringing in another familiar suspect: the golden angle.

## A Non-Fibonacci Example

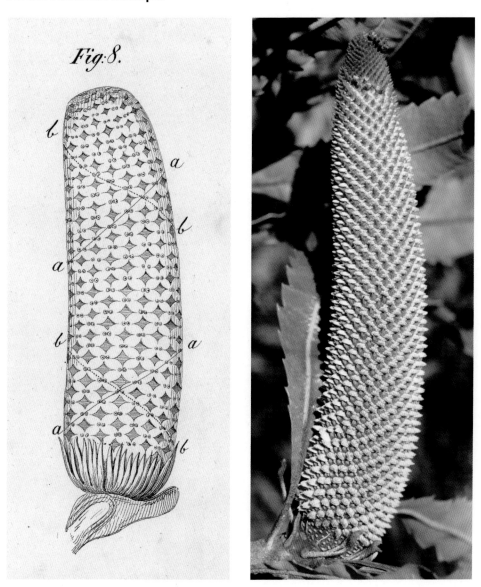

**FIG. 6.9** Here Braun provides a wonderful counterexample to Fibonacci: the flower spike of a *Banksia* plant from Australia, with eight a–a spirals and nine b–b spirals. Braun reads it as a 2/17 divergence. At right, *Banksia serrata* growing at the Royal Botanic Garden in Sydney.

## Tribute to Braun

**FIG. 6.10** This wonderfully detailed poster highlights the lifework of Braun, at a dinner honoring him in 1876, the year before he died. At top, the spiral illustrates his concept of the divergence angle—strangely, here a non-Fibonacci (7, 11) parastichy mode, despite the (5, 13) label.

Beneath the spirals is a drawing of a moss named for him, *Braunia sciuroides*. At right is one of his pinecones with scales carefully numbered, showing a vertical orthostichy, as in figure 6.4.

The microscopic views refer to Braun's work on cryptogams and eukaryotes, as well as his important contributions to cell theory. To the left of the menu is a *Pediastrum*, a genus of freshwater algae that can reproduce both sexually and asexually. At bottom are eukaryotes commonly known as "water molds," likely the *Saprolegnia prolifera* that Braun discussed in a paper.

At the poster's bottom corners are two plants that Braun studied: at left, an aquatic fern, *Marsilea*, which resembles a four-leaf clover; and at right, the unusual fern *Pilularia*.

The menu reveals that the eight-course dinner was a lavish affair, starting with Spring Soup, moving on to Lobster Pâté, Venison, and French Chicken, and ending with Swedish Pudding and Macerated Fruits. Of all the courses, only the asparagus had a connection to phyllotaxis (see recipe in chapter 21).

# Try Your Hand

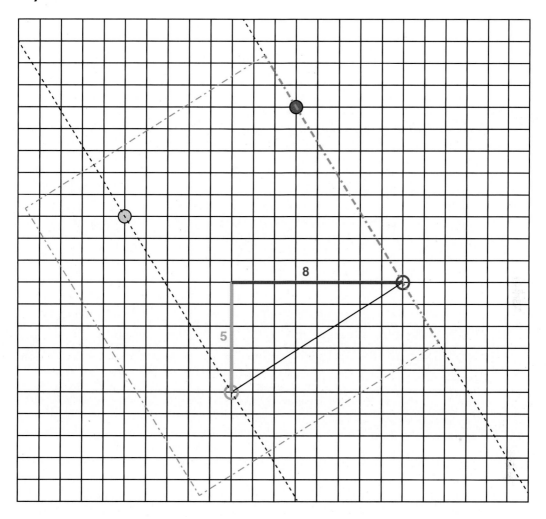

**FIG. 6.11** A simple stem diagram.

## Enter a Different Dimension

Follow the instructions here to make a simple cylindrical "stem" marked with a phyllotaxis pattern. The model helps conceptualize how a Fibonacci pattern looks when seen both on a flat plane ("unrolled") and on a cylinder.

1. Take a sheet of graph paper with large (¼″) squares. Choose a pair of Fibonacci parastichy numbers ($m$, $n$). In this example, we'll use (5, 8).
2. Mark a point near the bottom of the page, a bit left of center. Now count $m$ steps up from this point and then $n$ steps to the *right*. Mark a second point.
3. Next, starting from the first point, do the reverse: count $n$ steps up and $m$ steps to the *left*, and mark a third point.
4. And finally, do the same but starting from the *second* point, counting $n$ steps up and $m$ steps to the *left* again, marking where you land as the fourth point.
5. Draw lines connecting the two original points to the two new ones: these will be the edges of your cylinder.
6. Cut along the right line. On the left side, leave extra room for overlap before cutting. (Follow the blue lines in fig. 6.11).
7. Roll the paper so that the first two points meet and the two side lines overlap. You will now have a perfect ($m$, $n$) square phyllotaxis pattern. We will see this square pattern in the "unrolled" pineapple in figure 10.6, for example.

Don't tape the cylinder together. Instead unroll it again and ponder how Fibonacci phyllotaxis looks when 2D is converted to 3D—and back again!

Even more fun: using Braun's system, can you number the points on the cylinder, with parastichies that match your ($m$, $n$)? Start by labeling the first and second points you marked as 0. (Remember that they meet when the stem is rolled.)

### Label Florets by "Birth Order"

If you have a good picture of a flower whose spirals are clearly visible in both directions, you can label all the florets by order of appearance.

For practice, you can also photocopy the annotated dahlia in figure 6.13, enlarging it enough so you can easily fill in the floret numbers.

1. To start, draw the spirals in both directions and count them. Here we have 13 green clockwise spirals and 21 red counterclockwise ones.

**FIG. 6.12** From your starting point at 0, the floret numbers increase by 13 in the clockwise spirals, and by 21 in the counterclockwise spirals.

2. Now choose a floret near the outside and label it 0. Because it is growing near the outer rim, this floret will be one of the largest and oldest.
3. If there are $m$ clockwise spirals, starting at 0 you can label the next florets in the spiral by adding $m$ at each step. So the florets along this spiral will be $m$, then $2m$, then $3m$, etc. Here, with $m = 13$ clockwise spirals, the order will be 13, 26, 39, 52, . . .
4. Next, fill in the numbers starting at floret 0 for the spirals turning the other direction. These begin $n$, $2n$, $3n$, etc. In this example, with $n = 21$ counterclockwise spirals, the numbers are 21, 42, 63, . . .
5. Once you have labeled the two spirals starting from 0, you can fill in the rest of the florets. What should the floret to the right of 26 be labeled? Simply *subtract* 21—since you are moving *away* from the center. This means it should be labeled 5. Since it is a green spiral, you can continue to add 13 as you move toward the center, in order to get the remaining numbers for that spiral.
6. Keep going. Add $m$ or $n$ when moving toward the center of the flower and subtract $m$ or $n$ when moving away from the center, until you have labeled

**FIG. 6.13** Photocopy this dahlia image, enlarging it enough that you can easily draw in the missing numbers.

every floret. Check that no number repeats, a property that comes from $m$ and $n$ being relatively prime.

There are two emotional moments. The first is when you finally reach the number 1, which marks the first floret to appear approximately 137.5° away from point 0, where you began. (Take a moment to note where 2 and 3 fall in relation to each other.) And the second emotional moment is when you see the numbers cross—for example at floret 61—giving you the same result from both directions.

Note that this technique also works for the scales on a pinecone. Many phyllotaxis scientists used this approach to label scales or florets, including Auguste Bravais and Alan Turing.

CHAPTER 7

# Irrational Angles in a French Garden

> Brave
> soul,
> bravo
> Bravais, the
> crystallographer
> and divergence angle genius
> whose mind itself diverged and drifted far out to sea.

It's never a good feeling when, as you're racing to publish your paper, you find out that someone has just beaten you to the punch. That was how the Bravais brothers in France felt when they learned about the articles published in Germany by Schimper and Braun.

Fortunately, the two Bravais brothers, Louis (1800–1843) and Auguste (1811–1863), had taken a very different tack to solving the puzzle of Fibonacci phyllotaxis. In their foundational 1837 paper, they laid out concepts still used today, combining observations of real plants with mathematical rigor. If their article on plant geometry is notable for its clarity, it was nonetheless based on something irrational.

The two brothers used the relationship of Fibonacci numbers to the irrational golden ratio, seeing the golden angle as the essential organizing principle for most plants. Ultimately, their idea would stick a little too well, sometimes hampering future ways of imagining phyllotaxis. As for the two brothers themselves, the older one, Louis, a physician like his father, died young. But Auguste became a renowned explorer and physicist, whose intense life had a strange ending.

## Life of Auguste

If ever a scientist was built for the interdisciplinary field of phyllotaxis, it was Auguste Bravais. A true polymath, he wrote on a dizzying array of topics, from magnetism to seaweed to upside-down rainbows in the sky. Bravais was also a tough guy, a naval officer who, after charting the coast of sweltering Algeria, spent a freezing winter in Lapland collecting data on the northern lights. Best known for his mathematical models of crystals—still called Bravais lattices—he had a passion for finding intersections of numbers and nature.

"All my works," he wrote, "even those on pure geometry, were written with the aim of finding either a practical application, or of observing a natural phenomenon, in hopes of having the good luck to discover the explanation behind it."[1] His all-around brilliance, which brought him recognition at an early age, made his colleagues jealous.

Born in the mill town of Annonay, in southeast France, Auguste came from a family of plant lovers. His father, Victor, a physician, often took his children "botanizing"—hiking through the countryside collecting and identifying plants. In industrial Annonay, plants served a practical purpose, as dyes for leather, fabric, and paper. The town was also home to the Montgolfier brothers, the paper makers who invented the hot-air balloon.

**FIG. 7.1** The caption of this portrait in a French magazine calls him "Auguste Bravais, adventurer and scholar."

But Auguste was too restless and ambitious to remain in the provinces. At his Catholic high school in Paris, he dutifully studied classics, as his father wished—not opening until late at night the math books hidden in his trunk. His mathematical gifts would earn him a coveted spot at the École Polytechnique in Paris, the top French military-scientific university. After graduating at the top of his class in 1831, he joined the navy. The next year found him on the brig *Loiret*, charged with mapping an unexplored part of the coast of Algeria.

Despite the demands of his voyages, Auguste found time to collaborate with his brother, Louis, on their groundbreaking paper on phyllotaxis. Together, they published their "Essay on the Arrangement of Curvi-serial Leaves"[2] in 1837, despite a bitter last-minute disappointment. The Bravais brothers begin their essay with a note:

> We had just put the final touches on our paper about the symmetry of primordia in plants, when we learned of similar works by Misters Schimper and Braun, first from Dr. Martins' note in the 1833 *Archives de botanique*, and later from the original works of these German authors.[3]

Not to be discouraged, the Bravais brothers argued that their paper was better. They wrote: "Our results differed from theirs in formulating a broader generalization, as well as a law that these botanists either had not seen or thought they could ignore."[4]

Soon after publication, Auguste embarked on another adventure, setting sail for the Arctic on the ship *La Recherche* (The Research). He spent the entire winter in Lapland, at the northern tip of Norway, recording data on magnetism and the aurora borealis with half-frozen hands. While there, he observed the upside-down rainbow now known as Bravais' arc, formed by sunlight refracted through ice crystals. Some call it a smile in the sky.

Returning to France, Auguste obtained a professorship in astronomy in Lyon, where he became a veritable publishing machine. In 1848 came his most

**FIG. 7.2** In 1838, Auguste Bravais sailed on the ship *La Recherche* to the Arctic, where he observed the atmospheric halo now known as Bravais' arc.

famous breakthrough, in the field of crystallography. With the same love for systematic organization that marked his work on plants, Auguste found that the structures of crystals fit 14 mathematical models, later named the Bravais lattices.

After many honors and triumphs, then came downfall. Auguste's father died, and a few months later his only child succumbed to typhus. In the aftermath, Auguste, only 44 years old, catastrophically lost both his memory and his brilliance. Unable to work, he lived another seven years in utter misery. His contemporaries called him a "martyr to science," for it seemed that Auguste's brain had simply burned out from overuse. Perhaps the stress of life at sea and extremely hard work had proven overwhelming. Or perhaps he could not overcome his grief. We will never know.

Is it possible that studying phyllotaxis can be dangerous to your health?

## The Bravais Brothers' Fine Figures

**FIG. 7.3** Auguste Bravais was highly skilled at drafting, and his illustrations are dense with mathematical meaning. The brothers' figure 2 at top right shows scales like those on a pineapple. Knowing the numbering property from Braun, can you quickly find the number of spirals in both directions?

Their figure 1 at top left is their most important illustration, which they referred to many times. It shows a lattice whose divergence angle equals the golden angle of about 137.5°. The same figure illustrates basic concepts such as parastichies (lines joining leaves to their nearest neighbors) and the generative spiral (joining successive leaves 0, 1, 2, . . .).

At the center of the page, the circular figure recalls Braun's radial view (see fig. 6.7), with the notable difference that there are no straight curves: this is the sign of irrational divergence.

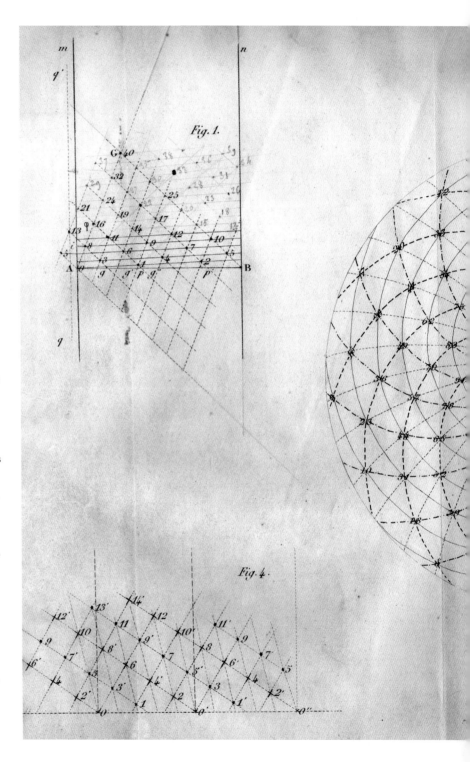

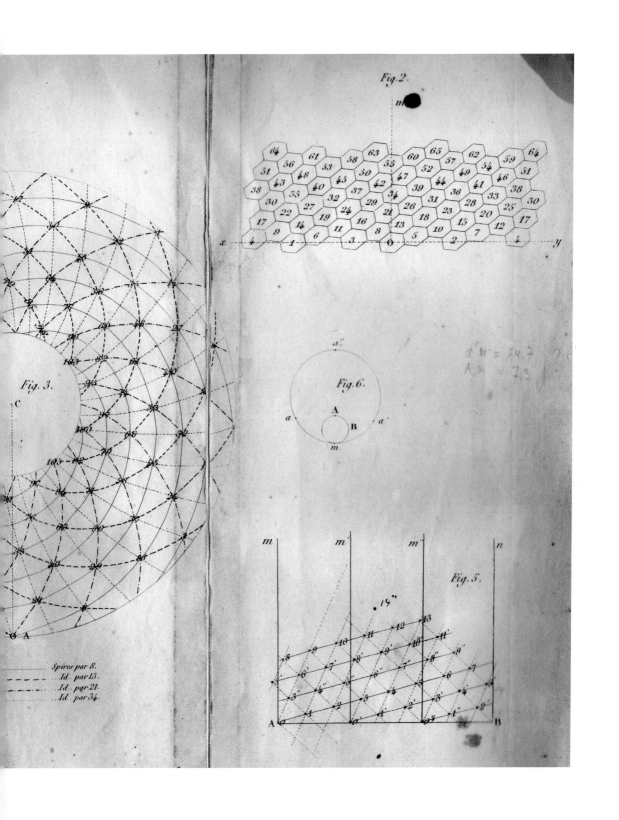

## Rational versus Irrational Angles

When the Bravais brothers learned about the work of Schimper and Braun, they scrambled before publishing their paper to distinguish their insights from those of the German botanists. Most crucially, they underscored the importance of a constant and irrational divergence angle.

In a beautiful interplay of theory and observation, Auguste and Louis argued that the divergence angle between one leaf and the next remains constant as a plant grows. (This applies to other plant organs as well, but to keep things simple, we'll use a leaf.) They determined that this angle is irrational, and more precisely that it is, in most plants, the golden angle. For the first time, they introduced the angle's exact value of $\frac{3-\sqrt{5}}{2}$ in units of turns. (This is about 137.508°, or 137°30′38″.)

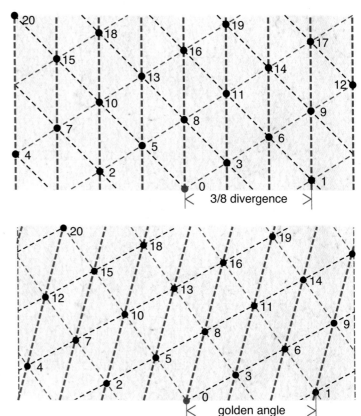

**FIG. 7.4** Rational versus irrational divergence angle. Which one is a better fit for real plants?

The rational divergence angle described by Schimper and Braun implies a vertical alignment of leaves—but when the Bravais brothers looked at real plants with spiral phyllotaxis, they did not observe perfectly vertical lines. Like Bonnet before them, they saw these lines slowly spiral up the stem.

Figure 7.4 at top represents a plant whose leaves line up in neat vertical columns, following the views of Schimper and Braun. The rational divergence angle is 3/8. In the bottom view, by tilting the vertical lines slightly, we get a plant with an irrational divergence angle, here equal to the golden angle.

## Remember This Lattice!

In setting out to prove that the divergence angle is both constant and irrational, the Bravais brothers hit upon some important results. They unraveled the connections between the divergence angle and parastichy numbers in lattices. Roughly 150 years later, these results would be repackaged in a slightly more general version and called the "Fundamental Theorem of

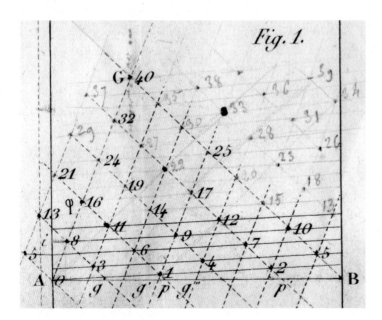

**FIG. 7.5** This is the Bravais brothers' all-important illustration, a lattice of points they labeled figure 1. In their large plate of illustrations, figure 7.3, it appears at top left. Note that an enthusiastic reader has numbered additional leaves in pencil (and drawn in some mysterious lines as well).

Phyllotaxis."[5] Sadly, however, the Bravais brothers did not receive the credit they deserved!

The story offers a perfect example of an odd phenomenon. In the field of phyllotaxis, well-meaning scientists have often repeated discoveries already made by others over the centuries.

Here, in the drawing they refer to many times, the brothers show a cylinder lattice unrolled onto a plane—a classic technique for analyzing plant spirals, which is still used today.[6] The thick vertical lines through A and B mark the same place on the cylinder, showing where its rolled edges meet. Going from A to B counts as one full turn around the cylinder.

In this sketch, all oblique lines represent spirals winding around the cylinder. The parastichies appear as dotted lines. There are 8 upward parastichies (note that the one through 5 and 13 overlaps at the left and right edges) and 5 downward parastichies.

As we move up the 8 upward parastichies, the leaf numbers increase by 8—just as we saw in Braun. Similarly, the leaf numbers rise by 5, following along the 5 downward parastichies. (From each point of the lattice, "all is the same," as the brothers put it.[7]) The generative spiral, which rises more slowly, is indicated by a solid line.

Now, how do we estimate the divergence angle—call this $\delta$, or delta—for this lattice? That is, how much of a turn do we make between one leaf and the next? Well, if leaf 5 were directly above B, we would be in the rational territory of Schimper and Braun. The generative spiral would pass through 5 leaves in 2 turns (as seen in the solid lines passing through leaves 1 through 5). So, from one leaf to the next, we would make 2/5 of a turn.

But here, the spiral doesn't *quite* reach 2 turns in 5 leaves. So the divergence angle $\delta$ is a little less than 2/5. We can play the same game with node 8: if it were directly above A, rather than slightly to the right of it, the spiral would make 3 turns in 8 leaves, giving us 3/8 of a turn per leaf. But here, the spiral makes a little *more* than 3 turns in 8 leaves, so $\delta$ is greater than 3/8.

In short, we can write:

$$\frac{3}{8} < \delta < \frac{2}{5}$$

Note that the four numbers satisfy this equation:

$$(8 \times 2) - (3 \times 5) = 1$$

This is no mere coincidence. Mathematician that he was, Auguste saw a theorem. He generalized it to any lattice whose parastichy numbers $(m, n)$ are relatively prime.[8] For such a lattice, he defined $\Delta_m$ as the number of turns you must make on the generative spiral for leaf $m$ to be nearly directly above leaf $0$.[9] And similarly for $\Delta_n$. From his example using $m = 5$ and $n = 8$, we obtain $\Delta_m = 3$ and $\Delta_n = 2$.

He then proved that the divergence angle $\delta$ must fall between the fractions $\frac{\Delta_m}{m}$ and $\frac{\Delta_n}{n}$, again as in the example above.

But Auguste did not stop there. He also gave his readers a useful shortcut, so that they could compute $\Delta_m$ and $\Delta_n$ without having to trace a lattice for every $(m, n)$ or count turns on the drawing. Once again, he turned to his figure 1 to demonstrate. He showed that (just as 2 and 3 relate to 5 and 8), $\Delta_m$ and $\Delta_n$ are the unique positive integers smaller than $m$ and $n$ such that

$$(n \times \Delta_m) - (m \times \Delta_n) = \pm 1$$

From this last identity, Auguste[10] saw how you can find leaf 1 from leaf 0 by going up $\Delta_m$ steps along the the parastichy by $n$ and down $\Delta_n$ steps along the parastichy by $m$.[11] He also saw that you could use the algebra of continued fractions to compute $\Delta_m$ and $\Delta_n$ for any coprime pair $(m, n)$. In a footnote, he even gave a crash course on continued fractions—so that his botanist readers could find $\Delta_m$ and $\Delta_n$ using algebra.[12] The beauty of the Bravais brothers' work is that they moved deftly between the abstract and the concrete, the specific and the general. They even showed how to use their newfound gadgets $\Delta_m$ and $\Delta_n$ to estimate, with great precision, the divergence angle on real plants.

## Fieldwork

The Bravais brothers measured hundreds, if not thousands, of real plants for their phyllotaxis article. Very likely, Louis did most of this painstaking work, which required keen eyes, thread, and a drawing compass. We know that Auguste was busy aboard his ship in France's new colony of Algeria, working out the mathematical parts of the phyllotaxis article as well as writing two unrelated papers that landed him a doctorate in mathematical sciences. However they divided the labor, the brothers worked well together, as their paper shows a constant interplay between observation, measurement, and theory.

FIG. 7.6 Yellow asphodel, a wildflower measured by the Bravais brothers. Do you see how each leaf is always a bit tilted away from the one below?

One of the plants they measured was a fragrant wildflower called yellow asphodel (*Asphodeline lutea*). The brothers weren't content to simply approximate its divergence angle using their theorem. As evidence for their golden angle concept, they wanted the most accurate measurements possible. And so they collected data in several ways for each plant before calculating an average.

Once again, they turn to their figure 1 for help. Recall that in this illustration, the divergence angle $\delta$ satisfies $\delta < \frac{2}{5}$. Auguste writes this as

$$5\,\delta = 2 + \delta_5$$

where $\delta_5$ is the divergence angle between leaf 0 and leaf 5. (They call this the "secondary divergence angle.") Now, how do they measure $\delta_5$ on a living plant? Here they describe one method:

> If we have in our hands a regular stem with clearly observable fibers, peeling back the bark if necessary, we can measure the circumference with a thread wound once or several times around. A compass gives us, between the two fibers, the chord of the arc we want.[13]

Using the radius of the stem and the length of the chord, they could then obtain the angle using trigonometry tables, as back then there were no calculators. Once you have $\delta_5$, you can obtain the divergence angle simply by dividing by 5:

$$\delta = \frac{2}{5} + \frac{\delta_5}{5}$$

And of course, what you can do with 5, you can do with 8 as well:

$$\delta = \frac{3}{8} + \frac{\delta_8}{8}$$

>  Divergence secondaire de la feuille 5 = — 31° 6'
>  de la feuille 8 = + 19 9'
>  de la feuille 13 = — 10° 37'
>  Divergence génératrice conclue... $\delta_1 =$ 137° 47'
>  137° 25'
>  137° 39'
>  Moyenne 137° 37'. (1)

**FIG. 7.7** Three measurements from a yellow asphodel, resulting in a mean divergence angle of 137°37'.

In pursuit of accuracy, the brothers go even further. For the asphodel, they also calculate $\delta = \frac{5}{13} + \frac{\delta_{13}}{13}$. (The brothers generalize this formula as $\delta = \frac{\Delta_n}{n} + \frac{\delta_n}{n} \ldots$) Next they take the mean of the results from these three measurements, which gives them a result very close to the golden angle of approximately 137.5°.

So ultimately, why did the Bravais brothers think that Schimper and Braun were wrong in insisting on a rational divergence angle? First of all, as noted above, they didn't see vertically aligned leaves in real plants. Also, they observed that some plants display two types of phyllotaxis on the same stem—which would mean the rational divergence angle would have to jump. But if the divergence angle is constant and equal to the golden angle, then it works for multiple spirals on a stem. (You would simply change the vertical pitch of the generative spiral.)

## Continued Fractions and the Golden Angle

Getting a bit more technical here, the brothers also found a mathematical road to the golden angle by using continued fractions. They started by observing that the parastichy numbers of most plants are pairs of adjacent Fibonacci numbers—often (3, 5), or (5, 8). The largest they saw was a sunflower (*Helianthus annus*), whose spirals were (89, 144).

According to their theorem, the *approximate* rational divergences for these phyllotaxis modes are

$$\frac{\Delta_2}{2}, \frac{\Delta_3}{3}, \frac{\Delta_5}{5}, \frac{\Delta_8}{8} \ldots$$

They also argue that the $\Delta$s also follow the Fibonacci sequence. So, for example, $\Delta_{13} = \Delta_5 + \Delta_8$. This is because if it takes about $\Delta_5 = 2$ turns to reach leaf 5, and about $\Delta_8 = 3$ turns to reach leaf 8, then it must take about $\Delta_5 + \Delta_8 = 5$ turns to reach leaf 13 (being $5 + 8$). Readers can verify this in the Bravais brothers' ever-useful figure 1. Each fraction is the mediant of the previous two, as Schimper had found, and like him, they could obtain the sequence of rational divergences:

$$\frac{1}{2}, \frac{1}{3}, \frac{2}{5}, \frac{3}{8}, \frac{5}{13}, \frac{8}{21}\ldots$$

These they view as successive approximations of the true divergence, and never the absolute state of the plant. And like Braun, they recognize these fractions as something mathematically elegant: successive truncations of the same infinite continued fraction. For example:

$$\frac{2}{5} = \cfrac{1}{2 + \cfrac{1}{2}}, \quad \frac{3}{8} = \cfrac{1}{2 + \cfrac{1}{1 + \cfrac{1}{2}}}, \quad \frac{5}{13} = \cfrac{1}{2 + \cfrac{1}{1 + \cfrac{1}{1 + \cfrac{1}{2}}}}, \quad \frac{8}{21} = \cfrac{1}{2 + \cfrac{1}{1 + \cfrac{1}{1 + \cfrac{1}{1 + \cfrac{1}{2}}}}}$$

All are truncations of this infinite continued fraction:

$$\cfrac{1}{2 + \cfrac{1}{1 + \cfrac{1}{1 + \cfrac{1}{1 + \ddots}}}}$$

Now the brothers took this idea two steps further. Using algebra, they noted, you can compute the value of the infinite continued fraction, "here equal to $\frac{3 - \sqrt{5}}{2}$, with $\sqrt{5}$ denoting the *irrational* quantity which, multiplied by itself, yields 5."[14]

Note how they italicized "irrational," a word dear to their hearts. This is of course the exact expression of what came to be called the golden angle, which they computed in degrees to be approximately

$$\frac{3 - \sqrt{5}}{2} \times 360 \approx 137°\ 30'28''$$

So this angle was, for the Bravais brothers, the key to the whole theory. In addition, but more as an aside, they made a connection to the golden ratio, stating that the golden angle splits the circle into an extreme and mean ratio, defined in a footnote.[15] This is all they said about the golden ratio in plants. And this is probably all there is to say.[16]

## Coda: The Bravais Lattices in Crystallography

The case of Auguste Bravais is highly unusual in one way: his study of phyllotaxis directly inspired his most famous work, a study of crystal structures. Plants gave him a vision of regularly placed elements. Using the generative spiral and a constant divergence angle, he was then able to work out the mathematics of plant spirals. It was essential that the same pattern could be seen in different ways, as Braun also noted, by following more horizontal or more vertical spirals.

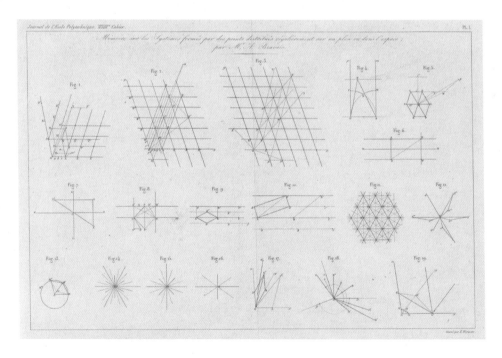

**FIG. 7.8** Applying his love for systematic organization, Auguste Bravais found that the structures of crystals fit 14 mathematical models—now called the Bravais lattices. Here, he lays them out in his important 1848 paper. Do you see in his figure 3, top center, a different way of drawing "spirals" in a familiar lattice?

**FIG. 7.9** This pinecone beautifully displays (8, 13) phyllotaxis.

These properties and their mathematical relations would also form the core of Auguste's work on crystals. More specifically, he translated the lattices of plants and their mathematical relationships into the various three-dimensional lattices formed by crystals. In this way, botanical investigations led to important discoveries in math and physics.

**Recap: The Era of Fixed Geometry**

Here we pause for a moment to look back over the first major phase of phyllotaxis discoveries, based on the idea that plants have fixed geometries. It's worth noting that the four scientists we've recently met—Schimper, Braun, and the Bravais brothers—started from the same observations of Fibonacci numbers but ended up in very different places.

- Schimper introduced the divergence angle, the most vertical spirals (orthostichies), and the generative spiral. He explained Fibonacci numbers as jumps from one rational value to the next, alternating between left and right.
- Braun introduced parastichies. He observed that even with a rational divergence angle, many spirals displaying Fibonacci numbers can be observed at the same time, using the property of addition and subtraction. He calculated the phyllotaxis of a pinecone using the most vertical set of spirals he could observe.
- The Bravais brothers followed a similar line of thinking, using various spirals all visible at the same time. But they linked all the Fibonacci numbers to a fixed irrational divergence angle, the golden angle. In addition, some of their work reads as theorems and proofs, representing a phyllotaxis first.

Certainly, all four of these scientists deserve credit for building the foundations of phyllotaxis. Still today, the fastest way to evaluate the phyllotaxis of a stem is by counting the number of leaves growing between two leaves that are nearly vertically aligned.[17] And the simplest way to number closely packed scales is to count the number of spirals. Those oh-so-perfect cylindrical lattice models have inspired many volumes, as has the beautiful mathematics they generate.

And of course, the Fibonacci sequence and the golden mean are now engraved, for better or worse, in the minds of many when thinking of plant patterns.

But none of our four scientists could offer a satisfying answer to this book's central question: Where do the Fibonacci numbers come from? The explanations of Schimper and Braun are fuzzy. The Bravais brothers merely shift the question to another problem, for why should the divergence angle be tied to the golden angle? For better answers, we must now shift to a new way of looking, based on dynamics. For this, first of all we need an expert on how plants grow, a self-taught German botanist by the name of Wilhelm Hofmeister.

## Try Your Hand

### Make Fibonacci Flowers

In this activity, you will see how the golden angle of divergence generates Fibonacci numbers of spirals, just as the Bravais brothers said.

To prepare, first photocopy the "pie tool" and circles onto a single sheet of paper. Cut out the tool, which measures the golden angle (actually a smidgen less, to account for the thickness of your pencil tip). The circles represent the different levels on the plant stem, with the central circle being the edge of the meristem.

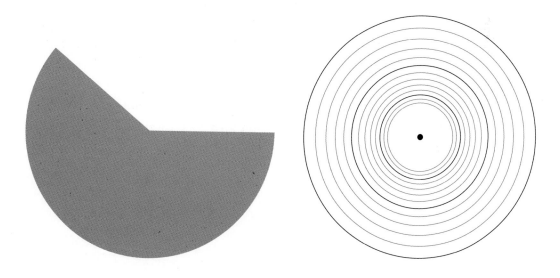

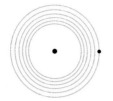

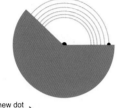

new dot
(on 2nd circle)

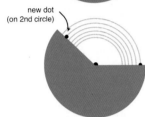

new dot
(on 3d circle)

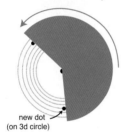

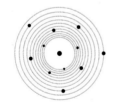

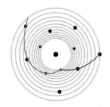

**DIRECTIONS**

1. Draw a black dot somewhere on the outer circle. This represents the center of a primordium, or leaf embryo.
2. Line up one edge of your pie tool, aligning the dot at the center with the dot you just drew. If you'd like, you can pin the tool to the center and rotate it around.
3. Now draw a new dot along the other edge of the pie tool, one circle in.
4. Turn your pie tool counterclockwise, to line it up with the dot you just drew. Now draw another dot along the opposite edge, one more circle in.
5. Repeat until you have a dot on each circle. Always turn the pie tool in the same direction, and be sure to keep track of the circle you are on at each step. Counting helps!
6. Now look for the spiral arms formed by the dots, winding in two directions. First, using a red marker, connect the spiral arms turning counterclockwise. Starting from the last dot you made, find the curving path of dots to the outermost circle, arcing to the left. (Hint: successive points should be separated by the same number of circles each time.)
7. Continue to draw red arms, using all the dots without using any dot more than once. (Note: be sure to start on the inner circle, not the black dot at the center.)
8. Next, using a green marker, connect the spiral arms turning clockwise. These will be longer arms, passing through three or four points.
9. How many spirals do you have of each color?

**Disclaimer!** The authors do **not** believe that plants place new primordia by measuring angles in this way, as the Bravais brothers seemed to think. Instead, the angle is a property of the primordia formation process (see chapters 13 and 14).

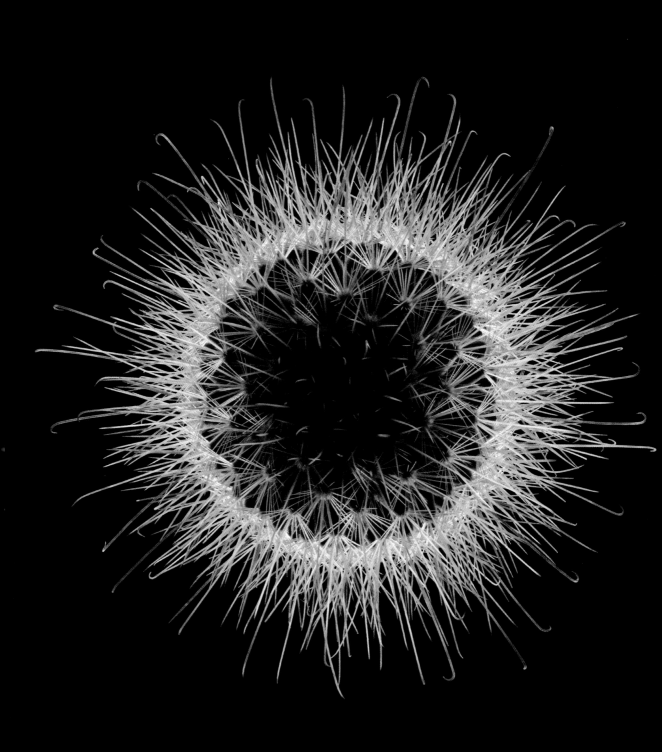

PART III

# What Did the Microscope Reveal?

# CHAPTER 8

# A Glimpse of the Growing Tip

    Self-
    taught
    skeptic
    Hofmeister
    wielded his sharp knife.
    He stripped all away to reveal
    the hidden meristem, where new plant cells greet the world.

Although the German botanist Wilhelm Hofmeister (1824–1877) did not work directly on plant spirals, he discovered a rule essential to solving the mystery of why Fibonacci numbers appear. Up to this point, our study of plant patterns has focused on analyzing spirals visible in full-grown plants. But Hofmeister, an independent thinker who was mostly self-taught, sought to understand these patterns by asking a very different question: How do plants grow?

Raised in Leipzig, Wilhelm left school at 15, taking a job in his father's music publishing business.[1] His dreams had nothing to do with violins, however. Thirsty for knowledge, he woke up at 4 a.m. every morning to gaze at minute details of plants, setting up his own independent lab. In 1851, when he was just 27, his dogged determination paid off. Hofmeister published his landmark paper describing the alternation of generations in mosses and ferns. He showed that

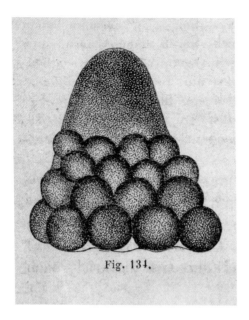 

**FIG. 8.1** At left, bumps that Wilhelm Hofmeister observed through his microscope. He wondered how and where they formed. At right, a tip of Romanesco broccoli. How do the spirals compare?

these plants have two entirely different life phases: one generation produces spores, while the other produces sperm and/or egg cells. This was totally eye-opening, given that his contemporaries did not even know that plants developed from egg cells.

Nearsighted from an early age, Wilhelm refused to wear glasses. While he couldn't see much at a distance, he had the hands of a surgeon and an extraordinary ability to focus on tiny plant organs. These skills were crucial to his studies of plant reproduction. "He could extract from the ovary a single ovule scarcely visible to the naked eye,"[2] recalled his student and biographer, Karl von Goebel. And then Hofmeister could, "even without using a dissecting microscope, detach the embryo-sac with a needle."[3]

To put his work into a larger context, it's important to note that Hofmeister rejected the spiritual philosophy that dominated nineteenth-century plant science. In Germany, the botanical writings of poet Johann Wolfgang von Goethe had spawned an influential movement called "idealistic morphology." Hofmeister dismissed it as nonsense. He belonged, as his biographer put it, "to those who keep science and poetry apart."[4] Working alone, Hofmeister discovered the alternation of generations based solely on his observations, uncolored by scientific fads of the era. His paper so impressed established botanists that (although initially not all agreed with him) the University of Rostock gave him an honorary doctorate.

Still, it was not until 1863, when Hofmeister was 39, that he was able to leave music publishing for good. He landed a job at prestigious Heidelberg University, as a full-time professor of botany. There, he was at last able to focus fully on his microscopy work, and he wasted no time. In 1868, he published his book *Allgemeine Morphologie der Gewächse* (General morphology of plants),[5] which would change phyllotaxis studies and botany forever.

FIG. 8.2 Wilhelm Hofmeister was photographed in 1872, a few years after publishing what would become known as "Hofmeister's rule." Self-taught, he was endlessly curious and energetic.

Once again, Hofmeister broke with the idealists. This time, he dismissed the work of Schimper, Braun, and the others who embraced the "spiral theory" of phyllotaxis—those who believed that leaves and other plant organs followed a predetermined spiral. Hofmeister wrote that the idea of primordia "proceeding in a screw-line or a spiral is not only an inappropriate hypothesis. It is an error."[6] Only by abandoning that concept, he wrote, could scientists truly begin to understand the predominant patterns in plants.

Dismissing Braun's spirals, Hofmeister instead based his work on the microscopic discoveries of Swiss botanist Carl von Nägeli. In 1845, Nägeli had described and named the *apical meristem*, the tiny region at the tip of a bud where plant organs first develop.[7] Hofmeister pursued this line of research under the microscope, observing how and where plant development occurred, using his unusual powers of observation.

His insights then led to a cornerstone of phyllotaxis now known as "Hofmeister's rule." This fundamental idea would underlie many phyllotaxis investigations that followed, as scientists began using physics to understand why Fibonacci spirals form in plants. As his biographer stated: "Hofmeister is the founder of the mechanistic interpretation, which afterwards was further developed, especially by [Simon] Schwendener,"[8] the Swiss scientist we will meet in the next chapter. It was not only Schwendener who would be influenced, however, but also every physicist who would approach the subject of plant patterns from this point on.

**Observing the Bumps**

Although Hofmeister didn't work directly on phyllotaxis, he was among the first scientists to describe in detail the apical meristem, where Fibonacci spirals originate. The meristem contains new cells that all appear alike, as they have not yet differentiated into distinct organs like stems or leaves. The term "apical" refers to the very tip (apex) of the plant. Being a self-taught scientist who disliked pretension, Hofmeister simply described what he saw under the microscope, trying to find the simplest possible way to explain his observations. His approach led to results that were both important and enduring.

Starting at the tip of the tip, Hofmeister described a central embryonic zone where very little growth occurs. Just beyond it, Hofmeister observed a zone where little bumps appear—the primordia. These then develop into the various botanical organs, such as stems and leaves. None other than Johann Wolfgang von Goethe, the poet-scientist, had already proposed the common origin of all plant organs, writing in his 1790 book *Metamorphosis of Plants* that "all is leaf." But Hofmeister gave this idea powerful validation when he observed how all plant organs begin as the same little bumps at the border of the bud tip.

**FIG. 8.3** Here, Hofmeister has drawn what he saw under the microscope: an apical meristem with the bumps of new plant organs forming. Using the numbering property, can you see how many spirals there are?

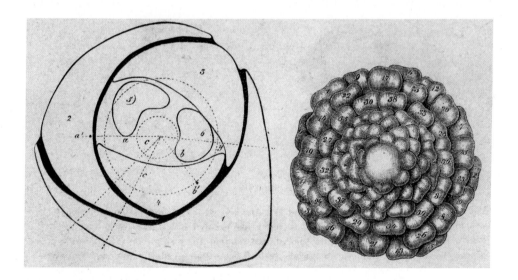

**FIG. 8.4** At left, Hofmeister has drawn a bud cut just at the height of the apical meristem, indicated by the central circle. At right is a drawing of a similar meristem viewed from above, this time when the plant is flowering, showing many more primordia. Could both grow in the same way?

In the left-hand illustration in figure 8.4, Hofmeister has sliced through a bud exactly at the height of the apical meristem. The newest primordia (closest to the central circle) are shaped like beans. The next oldest primordia, having grown larger, now look like sharp quarter moons. This drawing would become iconic for scientists studying phyllotaxis in the bud, using a circle for the meristem with radial lines emerging from the center.

Again based on his observations, Hofmeister described how these bumps expand. Without any preconceptions—and not assuming any in the reader—he described the evolution of their shapes: the primordia first expand in length (forming the leaf's central vein) and then expand sideways (forming the leaf's margin), without changing much in thickness.

**Hofmeister's Rule**

Using a similar approach, Hofmeister also described the spacing of the region where new plant organs appear. While the central zone is nearly empty, the primordia farther away are tightly packed together. In this way, each new primordium appears between two previous ones. Relying on these observations, Hofmeister speculated about a cause. Here he turned his observation of a densely packed bud into a *stacking principle*, now known as Hofmeister's rule.

As he stated it in his 1868 book:

> New leaves (or lateral axes) appear in the spaces above the circumference of the still-growing stem-tip (or stem-zone) that are farthest from the lateral margins of the axes of the nearest leaves already present.[9]

Or to put it in simpler terms:

**A new primordium develops where the most space is available.**

This rule, which we will return to again and again in this book, not only describes where the new primordium is but also prescribes where the next one should form.

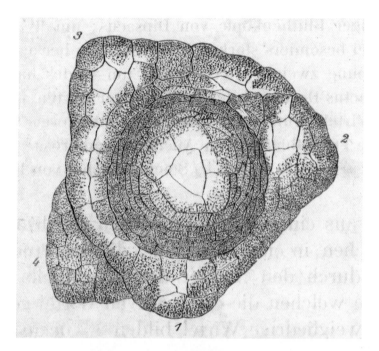

**FIG. 8.5** Here, Hofmeister has drawn an apical meristem viewed from above, with the individual cells shown and the surface beautifully rendered. The drawing reveals nearly as much detail as the latest images using electronic microscopy, like the one in figure 8.6.

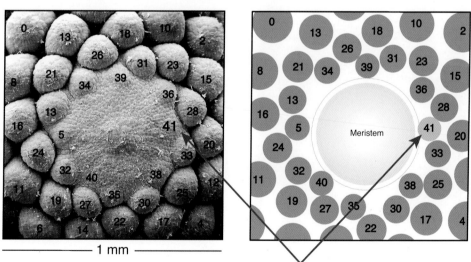

The new primordium forms in the least crowded space at the edge of the meristem.

**FIG. 8.6** In these two images, you can compare the apical meristem of a real plant with a computer-generated model that applied Hofmeister's rule. Where would the next new primordium appear?

In figure 8.6 at left, the scanning electron micrograph shows new needle primordia forming at the tip of a branch from a Norway spruce (*Picea abies*). The newest primordium is numbered 41, with the others numbered sequentially according to age. Observe how, as they grow, the primordia are displaced radially away from the center of the circular meristem. At right, the computer model shows the configuration of primordia that appears when Hofmeister's rule is applied. Note the remarkable similarity between the real plant and the computer model. Interestingly, the pattern scales up as the plant grows, remaining unchanged as the primordia become mature plant organs.[10]

Hofmeister did not push beyond his rule and never tried to explain how this stacking principle could lead to the patterns he observed. But his was a fundamentally new approach, focusing on the continuous development of the plant to explain its final form. He also made a major leap in recognizing that from a simple shape with simple growth rules, complex shapes can be obtained. This concept—related to what is now called dynamical systems and complexity theory—would later be expanded upon by Schwendener and then by Alan Turing. (As we will see in chapter 11, Turing was not only an inventor of the computer but also a scientist fascinated by the patterns in daisies and sunflowers.)

Today scientists use Hofmeister's dynamical approach to investigate many developmental processes, and so his work has remained relevant. In a paper titled "The Genius of Wilhelm Hofmeister," biologists Donald Kaplan and Todd Cooke state that although Hofmeister never became famous, he "stands as one of the true giants in the history of biology and belongs in the same pantheon as Darwin and Mendel."[11]

Moving on from Hofmeister in Germany, next we will explore the ideas of the Swiss botanist Simon Schwendener, another original thinker whose insightful interdisciplinary work has been underappreciated. Note that both these scientists wrote in German, and that neither managed to get many of his works translated into English. This helps explain why, in the twentieth century, their writings on phyllotaxis fell into obscurity.

## Coda: Hofmeister's Crystal Ball?

In many ways, Hofmeister's observations were ahead of their time. Specifically, his drawings in figure 8.7 foreshadow work on early computer L-systems and close-packing (discussed in chapters 13 and 17 of this book, respectively).

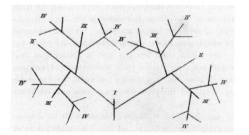 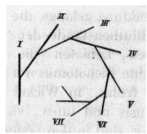 

**FIG. 8.7** Here, Hofmeister has sketched three schematic diagrams of possible architecture in plants. With these, he is a century ahead of his time. Can you guess the development rule leading to the three different shapes?

These sketches might appear simple, but they are actually sophisticated precursors of L-systems used in early computer simulations of phyllotaxis, with identical elements repeating themselves in a particular order.

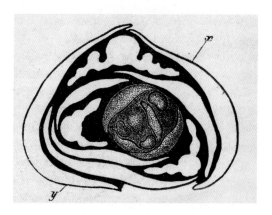 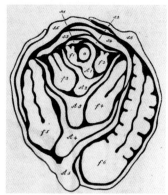

**FIG. 8.8** Hofmeister's drawings reveal the complex packing of leaves inside buds. Actually, the drawings show both true leaves and protective pseudoleaves. Do you see the difference? Can you find the phyllotaxis pattern of the true leaves, labeled $f$?

Additional drawings by Hofmeister (see fig. 8.8), showing the complex development of leaves inside buds, are precursors to studies of close-packing in the formation of leaves. They show cross sections of the large buds typical of oak, birch, and other trees. At left, the apical meristem is drawn in detail. Around it are primordia that have developed into leaves with contoured shapes, surrounded by secondary protective leaflets (stipules). At right, growth of the central veins has induced the young leaves to fold in two.

CHAPTER 9

# Biomechanics under the Lens

> Swiss
> sage
> Simon
> Schwendener
> says: the microscope
> shines light on biomechanics.
> He lived alone but stacked his disks tightly together.

The son of a farmer, Swiss scientist Simon Schwendener (1829–1919) made his way through sheer grit. He was so poor that he had to choose between marriage and botany. Botany won out. Not until he was 38 did he have a stable enough income to support a family, and by then he felt it was too late. Reflecting back on his life, he wrote, "Fighting for science I have grown old, but I was successful."[1]

A botanist by training, Schwendener described his research as "physico-mathematical." Phyllotaxis was a perfect fit for his wide-ranging mind, standing as it does at the intersection of biology, math, and physics. He would peer through his microscope at real plant cells and then make leaps into brilliant abstractions. How exactly did he help explain why Fibonacci numbers appear in plants? Most importantly, he was the first to create a model for leaf placement using circles, an approach now known as disk stacking. Remarkably, this model is still used today,

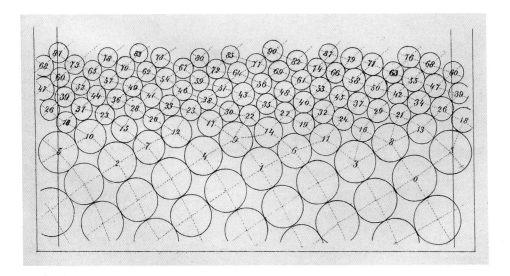

**FIG. 9.1** In 1878, Simon Schwendener introduced disk stacking, now used in computer simulations of phyllotaxis. He also revealed transitions between patterns, obtained when reducing the size of disks. How many spirals can you count before and after the transition? Do you see anything unusual there?

with the added firepower of computer simulations. With his model, Schwendener illustrated transitions between successive spirals in Fibonacci phyllotaxis (see fig. 9.1), something that Dutch botanist Gerrit van Iterson and the authors of this book would pursue more systematically. In addition, Schwendener asked prescient questions: How do new leaves form over generations? How do we understand plants with irregular phyllotaxis? How do plant patterns deform under pressure?

As a biologist, Schwendener's claim to fame was a stunning discovery involving lichen. In 1867, he argued that lichens are not distinct organisms but instead two life-forms living together: fungi and algae. His then-radical idea was based on his observations through the microscope, for he was a skilled microscopist who would later coauthor a classic manual, *The Microscope in Theory and Practice* (1877).[2] Although within lichens he described a master-slave relationship that turned out to be inaccurate—in fact the fungi and algae live cooperatively—his observations were key to unlocking the new concept of symbiosis.

**FIG. 9.2** Schwendener discovered that lichens are actually two life-forms living together, fungi and algae. This lichen, *Evernia prunastri*, grew in the garden of Stéphane's great-grandfather, entomologist Rémy Perrier. Does it look like a mushroom? Or like algae?

**FIG. 9.3** Portrait of Simon Schwendener.

In 1874 came another landmark, a paper that would launch an entire scientific discipline: plant biomechanics. Schwendener, armed with his microscope, had little patience for botany that merely described stems and stamens. Was Aristotle first to apply math and physics to plants? We will never know, as his writings are lost. So it is that, according to specialists, "plant biomechanics became a clearly defined field of study upon the publication of Simon Schwendener's seminal book,"[3] *The Mechanical Principle of the Anatomy of Monocotyledons*[4] (as it would be in English). His study investigates the relationship in plants between structural tissue and physical stress responses. In applying his physico-mathematical method, Schwendener believed he was taking a sharp turn away from traditional botany's descriptive approach.

Four years later, in 1878, Schwendener published his magnum opus on phyllotaxis, taking a similar approach. This was his *Mechanische Theorie der*

*Blattstellungen* (Mechanical theory of leaf arrangement). Strangely, the work has been all but forgotten: the entry on Schwendener in the *Complete Dictionary of Scientific Biography* does not even give it a mention. Like Hofmeister before him, Schwendener laid out a mechanistic approach that was fundamentally dynamical. In his phyllotaxis paper, he returned to ideas from his doctoral thesis, which began with a powerful sentence: "All organic life is based on the game of continuous formation and deformation."[5]

It's interesting to note that Schwendener appears to have had a feminist side. He had at least one female postdoc, Grace D. Chester, who worked with him "on stomata of petals and anthers,"[6] according to a biographical sketch of Schwendener by Rosemarie Honegger. Three of the five early female members of the German Botanical Society were recommended by him. This was unusual at a time when, as Honegger noted, "other science professors at the University of Berlin, upon spotting a female student in the audience, personally escorted the 'respected Miss' (*gädiges Fräulein*) out of the lecture hall before starting their lecture."[7]

Why has Schwendener's phyllotaxis work been so neglected? It is true that his dense German text can be daunting. But Chris, who took on the challenge of analyzing it, came away a fan. "Schwendener is much bigger and a much greater influence than people think," he says. "It seems unfair that Gerrit van Iterson got all the credit for his ideas, many of them essential to the history of phyllotaxis!" (Van Iterson did at least praise Schwendener's insights, as we will see in the next chapter.)

Although Schwendener revealed little about his emotional life in the brief autobiographical sketch he wrote at the end of his life, he did occasionally write poetry. The following stanza from his poem "Confessions" was quoted in a lengthy obituary, published following his death in 1919 at age 90:

> This researcher's main concern
> Was how he could some money earn
> And so his daily bread obtain;
> Or else his lucky star would wane.[8]

## Introducing Disk Stacking

It was still new, in the 1870s: using a microscope to study botanical processes. Wilhelm Hofmeister and Schwendener were both pioneers in this regard, "going small" to understand plant mechanisms.[9] Observations of primordia under the microscope formed the basis of Schwendener's phyllotaxis treatise. He saw that new leaves did not form in a perfectly orderly way but instead showed irregularities in size and location. For Schwendener, this was evidence that they did not follow preordained spirals. Leaving fixed geometry behind, he embraced the approach of Hofmeister, who had published his crucial phyllotaxis rule a decade before.

It was Hofmeister who, Schwendener wrote, "introduced a mechanistic element for the first time." With Hofmeister's rule—that each new primordium forms where there is room for it—the generative spiral was out of the picture. "Now we only have to deal with dots," Schwendener wrote, "or, if we prefer, with figures that enter into certain relationships with others."[10]

Freed from the generative spiral, he made a conceptual leap. In his 1878 book, Schwendener proposed the first mathematical model of phyllotaxis pattern *formation*. Based on his microscope observations, he added postulates to Hofmeister's rule. First, primordia size remains nearly constant—independent of meristem circumference—for any given type of plant organ. And second, primordia grow as hemispherical bumps that eventually become tangent to one another.[11]

Then Schwendener began, on paper, stacking primordia one by one on the surface of a cylinder. Following Hofmeister's rule, he placed each primordium at the lowest place available without overlap. He noted that when he varied the shape of the primordia—using circles, ellipses, or "folioid" leaf shapes—this did not qualitatively change his analysis. Therefore, he reasoned, he might as well represent primordia as simple circular disks. And so disk stacking was born.[12]

Today, scientists use Schwendener's same basic technique: they stack up disks one by one on the surface of a cylinder, following the rule that the new disk must be added in the lowest possible place above the existing disks, without overlap. With this approach, each disk is tangent to at least two disks below it. Systematic computer simulations make the model that much more powerful. "As simple as it is," Chris notes, "this model exhibits plenty of geometric and dynamical richness, and it helps explain the predominance of Fibonacci phyllotaxis." We will see this in more detail in chapter 14, which presents the powerful approach called "zigzag fronts."

## Generations of a Disk "Family"

Using his disk model, Schwendener investigated how primordium placement changes as a plant grows—applying his postulates along with Hofmeister's rule. He focused not only on regular formations but also on more unusual occurrences, showing generations of primordia that some scientists refer to as grandparents, parents, and children.

In figure 9.4, Schwendener's top drawing shows two "children" on the left side of the "parent" (largest) ellipse, and one "child" on the right side. This formation is very common in plants, occurring in most moderately fast transitions. The middle figure, with two children on either side of the parent, occurs when there are more rapid regime changes, as in the teasel plant in figure 9.9. The bottom figure depicts a rarer case, with quite dramatic changes in the relative size of primordia. The effect of this dramatic change will be considered in chapter 14.

None of these three cases occur in regular lattice structures. Instead, they represent an

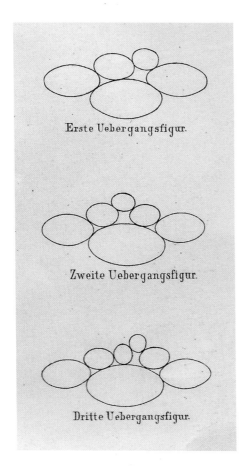

**FIG. 9.4** In these drawings, Schwendener models generations of primordia, with "parents" and "children." He also shows three basic pattern transitions the primordia can generate as the plant grows.

alphabet of how, as primordia size changes relative to the stem, stacked primordia induce what scientists call transitions. Transitions are places where the number of parastichies increases or decreases as a plant grows (see fig. 9.1 for a good example). If you're looking at seeds on a strawberry, for example, you can trace where a spiral leads to a dead end. For scientists, these unusual patterns have proven challenging to understand.

## Beyond Fibonacci

Using his disk stacking model, Schwendener could not only represent "normal" Fibonacci patterns but also explore the frontiers where these patterns fail. He examined several plants with unusual transitions, including members of the Araceae family like the peace lily (*Spathiphyllum cochlearispathum*, fig. 9.5), along with magnolias and teasels, the latter bearing purple flowers that resemble thistles.

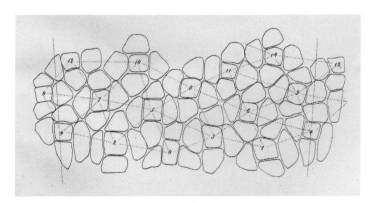

**FIG. 9.5** At left, using a camera lucida, Schwendener has drawn an unrolled flower spike of *Anthurium*. Like all members of the Araceae family, it bears a spike (or spadix) whose base interrupts the usual pattern of florets. The *Anthurium obtusum* at right and the peace lily below also belong to this family.

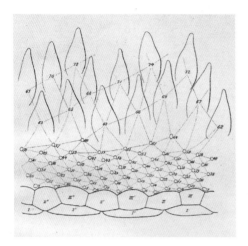 

**FIG. 9.6** At left, looking at another atypical case, Schwendener has drawn a magnolia flower unrolled on a plane. At right, Stéphane's magnolia photograph shows similar abrupt transitions.

In figure 9.6, Schwendener has drawn lines indicating where he has traced the parastichies of this magnolia inflorescence. There is an abrupt transition from a (3, 3) whorled pattern in the lower section, to an (8, 5) Fibonacci spiral pattern.[13] (Chris notes that he and his colleagues have not yet been able to reproduce this pattern using disk stacking on a computer. They might have to change to ellipses.)

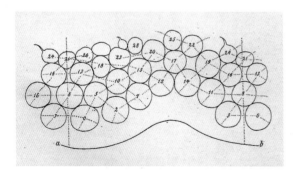 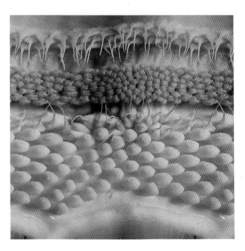

**FIG. 9.7** At left, Schwendener has sketched another unusual case, the flower spike of a cuckoo-pint (*Arum maculatum*). At right, the unrolled spike of a similar plant (*Arum italicum*) reveals a non-Fibonacci pattern we will revisit in chapter 14.

## Using a Camera Lucida

Why do Schwendener's drawings look so artless? He was sketching outlines of what he saw under the microscope using a simple projecting device called a camera lucida. The device offered an affordable alternative to microscope photography, which was at the time prohibitively difficult and expensive. It also enabled him to unroll, on paper, patterns he observed on plant stems. While the concept for a camera lucida had been proposed back in the seventeenth century by Johannes Kepler, the device was not manufactured until two centuries later, primarily for Schwendener's exact purpose: drawing specimens seen through a microscope.

To use a camera lucida, the scientist would attach the device to the top of the microscope tube. A prism (later optical mirrors) would allow the scientist to see a transparent "ghost" image of a specimen superimposed on the drawing paper. The scientist could then sketch the specimen's contours by tracing this shadow image. Success required practice, and the image, as Schwendener pointed out in his microscope handbook, was prone to distortion.

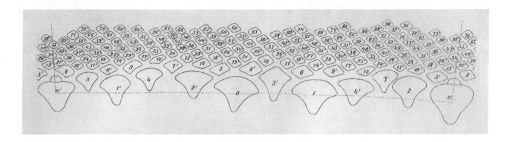

**FIG. 9.8** Again using a camera lucida, Schwendener has painstakingly sketched the unrolled primordia of a teasel (*Dipsacus*) flower head. Note how the parent primordium 0 has two children on the left, 2′ and 5; and two on the right, 8′ and 3′. Is the final pattern nonetheless Fibonacci?

Chris notes that the phyllotaxis in Schwendener's teasel drawing is *bijugate*, meaning the parastichy numbers are double the usual Fibonacci numbers: $(4, 6) = 2 \times (2, 3)$ at the bottom; and $(16, 26) = 2 \times (8, 13)$ at the top. As he did for his magnolia drawing in figure 9.6, Schwendener has indicated primordia belonging to the same whorl using 0, 0′, then 1, 1′, and so on.

**FIG. 9.9** A teasel. Would you dare to draw every floret, as Schwendener did?

## Deformation under Pressure?

Recall that Schwendener had long been interested in what he called "the game of continuous formation and deformation." But why did he begin his phyllotaxis treatise not with formation but with *deformation*? Aware that deformation logically comes after formation, he still put it first in his book: "In presenting my investigations, it seemed to me best to start with this latter process, for the simple reason that it can be given a strictly mathematical treatment, and therefore provides a secure basis for many further questions."[14] In his research, he conceived of mechanical experiments using cardboard tubes and also drew simple models of deformation using a ruler and compass.

One situation that particularly fascinated Schwendener was when a stem grows faster in thickness than in length. What happens then to the configuration of organs around the stem? Assuming first that the organs keep their original shape and size during this process, he modeled the situation using disk lattices.[15] Schwendener observed that when the lateral expansion of the stem is faster than the vertical expansion, the lattice is stretched sideways.[16] This deformation, with the disks remaining in contact, looks the same as if the lattice were uniformly compressed from above. He showed how, in the process, a lattice can go from (3, 5) to (5, 8) phyllotaxis and, by extension, through the Fibonacci sequence. Schwendener's ideas about compression and deformation would play a significant role in phyllotaxis theory—earning the name "contact pressure."[17]

Chris spent a long time working through Schwendener's little-known phyllotaxis treatise, and he came away with deep appreciation for Schwendener's accomplishments. "To me, Schwendener's paper marks the birth of the modern scientific view of phyllotaxis," Chris notes. "He relied on microscope observations, he looked for mechanistic principles of pattern formation, and he developed a model to go along with them. The way he moved back and forth between observation and theory is truly remarkable. The few modern authors that cite him tend to reduce his work to its first part, on deformation. But there is so much more to it!"

And as the cherry on top, Schwendener drew a zigzagging curve showing what happens to a disk when the lattice is under continuous deformation. That curve hints at the famous Van Iterson diagram we will meet in the next chapter.

**FIG. 9.10** Schwendener shows what happens to disks and parastichies when a stem grows faster in thickness than in length. The result is the deformation of a (3, 5) disk lattice into a (5, 8), moving down from the top of the page. Comparing the three lattices, do you see a difference in local shape?

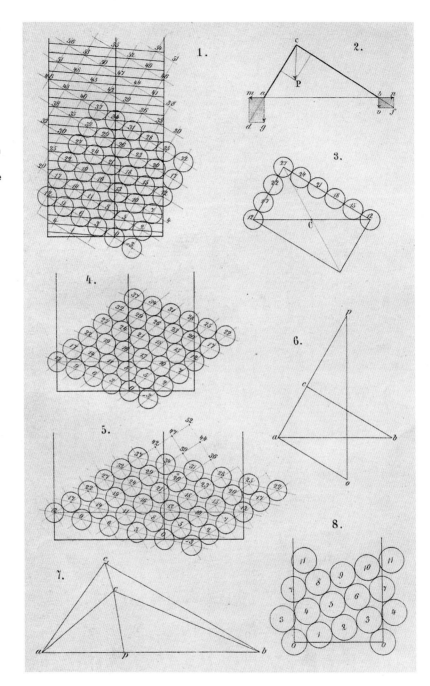

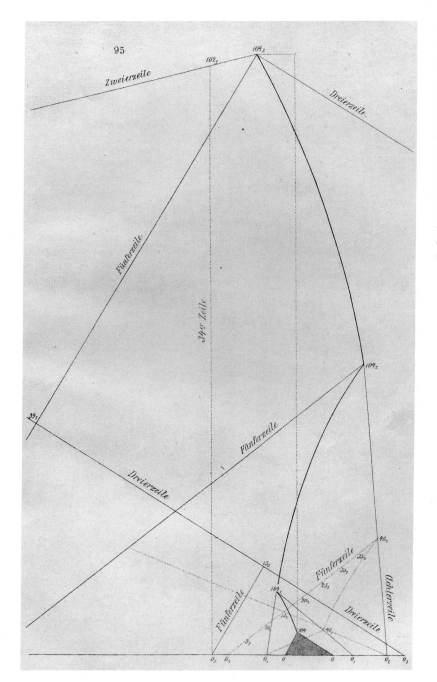

**FIG. 9.11** Here Schwendener has drawn the zigzagging curve that a disk center follows as the lattice expands in width. He claimed to see this in the leaf buds of a Spanish fir (*Abies pinsapo*) over the course of its growing season, as its phyllotaxis changed from (3, 5) to (8, 13). The result is a precursor to the Van Iterson diagram, coming up next.

# CHAPTER 10

# A Critical Tree on Graph Paper

> Wax
> balls
> led to
> his graphic
> revelations. It
> seems fitting that Van Iterson's
> diagram will so often be reiterated.

Pencils, graph paper, and disks made from beeswax. Using the simplest of materials, Dutch botanist Gerrit van Iterson Jr. (1878–1972) conducted science of enduring sophistication. His calculations led to one of the most iconic graphs in the field of phyllotaxis, now known as the Van Iterson diagram. The tree of possible solutions he published in 1907 serves as the phyllotaxis gold standard: all new models must pass the Van Iterson test, generally matching his diagram.

Fundamentally, Van Iterson's work builds on the step-by-step construction we saw with Hofmeister, while embracing Bravais's and Schwendener's mathematical approach. As the last of our scientists to conduct research without a computer, Van Iterson succeeded in turning disk packing into his beautiful mathematical diagram. Will it be enough to explain Fibonacci numbers in plants?

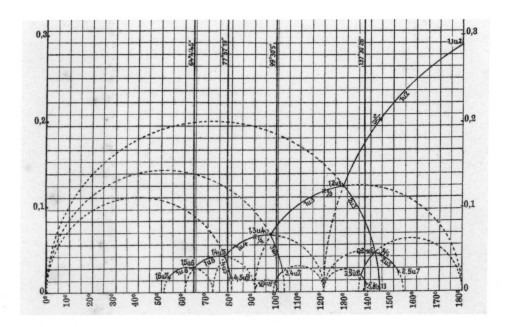

**FIG. 10.1** The Van Iterson diagram will play a starring role in this book. Do these arcs of circles replace Schimper and Braun's fraction tables? Can you guess how to follow the arcs to arrive at the golden mean—marked by the 137°30′28″ vertical line—at the bottom?

Van Iterson spent most of his life in Delft, a city with a long and distinguished history of scientific research. As a graduate student, he worked under the famed microbiologist Martinus Beijerinck, who, while studying tobacco leaves, in 1898 coined the word "virus." (The term comes from the Latin for "slimy liquid" or "poison.") Delft was also where, in 1673, Antonie van Leeuwenhoek first observed microorganisms in water, using a homemade microscope and the excellent lenses he ground himself.

Like many scientists drawn to phyllotaxis, Van Iterson was curious about virtually everything. Much of his career was devoted to practical botanical matters like improving rubber production in Indonesia. But according to his student Adrianus Meeuse, Van Iterson most loved exploring "the borderland between biology, mathematics, physics and chemistry."[1] With his wide-ranging intellect, Van Iterson often drew comparisons to his good friend D'Arcy Wentworth

**FIG. 10.2** Gerrit van Iterson at work. What do you suppose all the bottles were for?

Thompson, the Scottish mathematician who wrote the classic book *On Growth and Form* (for more on him, see chapter 17). Now and again, Thompson would make the trip from Dundee to Delft, visiting Van Iterson at home.

Van Iterson started out studying chemical engineering, but Beijerinck encouraged him to switch to phyllotaxis for his PhD. Van Iterson's doctoral thesis was an extraordinarily comprehensive study that, in 1907, would be published as a weighty book: *Mathematical and Microscopic-Anatomical Studies of Phyllotaxis*.[2] According to an article in *Nature*, the work "far exceeded in size, plan and contents everything normally offered as a thesis, and was the most convincing proof of the high scientific talents of the author."[3]

At Delft University of Technology, Van Iterson held the botany department's first chair in microscopical anatomy and also founded the university's botanical garden. He lectured with erudition on a dizzying range of topics, his student Meeuse recalled, "ranging from biostatistics and genetics to wood, fiber and paper microscopy; cell wall structure; X-ray diffraction of organic substances; biological membranes; vegetable oils; starch and starch products; economic botany; plant physiology and biochemistry."[4] Another account described Van Iterson as "careful and straightforward in conversation, and a good listener with a critical but warm sense of humor and unbounded energy."[5]

After teaching at Delft for three decades, at last in retirement he could return to the passion of his youth: phyllotaxis. At the age of 87, he published a new treatise, calling it part 1.[6] By way of fun facts, his son Frederik van Iterson was a prominent mechanical engineer, known for his iconic hourglass design of cooling towers for power plants.

## Beautiful Branches

In tackling the topic of phyllotaxis as a graduate student, Van Iterson was unusually thorough. Many scientists new to the field simply dive in, ignoring earlier writings or being unable to read them. Van Iterson, by contrast, devoured everything published on the subject. Like many Dutch citizens, he had the advantage of being a polyglot. (The town of Roermond, where he grew up, sits on a narrow strip between Belgium and Germany.) This meant Van Iterson could read the French of the Bravais brothers as well as the German of Schimper, Braun, Hofmeister, and Schwendener.

Not only that, he could read English. This brought Van Iterson to the mathematical phyllotaxis investigations of British botanist and illustrator Arthur Harry Church, whose exquisite artwork appears below. (We won't delve into Church's phyllotaxis story, as he had some very odd ideas, but Van Iterson nonetheless found his logarithmic spirals compelling.)

Concerning botany, Van Iterson had many insightful observations. He believed his theoretical work to be more important, however, and so real plants pretty much fell out of the picture. Building on the work of Schwendener, Van Iterson examined all disk lattices fulfilling the condition of the disks being in contact. He carried out this study more systematically than Schwendener by looking at a variety of different surfaces, including planes, cones, and cylinders. In each case, the results were beautiful treelike diagrams—the Van Iterson diagrams—that display a self-similar structure.

Van Iterson then tried to understand the formation and evolution of these patterns when the main parameter—disk size relative to stem size—varies over time as disks are stacked one by one. In addition, he conducted simple physical experiments in disk packing, using wax disks and wooden balls.

## Math behind the Famous Tree

Like many who have worked on the problem of phyllotaxis, Van Iterson began with the math, which is less speculative. Following in Schwendener's footsteps, Van Iterson first looked at regular packings, shown as lattices of disks on a cylinder.

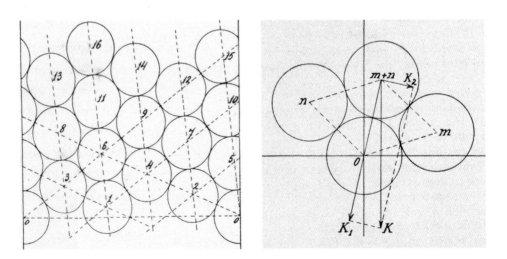

**FIG. 10.3** Here, Van Iterson identifies relationships leading to his tree. At left, the lattice shows disks 2, 3, and 5 in contact with disk 0, corresponding to the parastichy numbers. At right, his detail view shows a similar but general case, centered at disk 0. Does the left-hand pattern follow Braun or Bravais?

In figure 10.3, Van Iterson shows how the number of parastichies turning each direction, $m$ and $n$, relates to other parameters. These relationships will lead to his famous tree. As we have seen before, each disk represents a primordium (leaf or other plant organ). He starts by determining the initial condition of contact, putting the first three disks into place. The first disk at the bottom is labeled 0. He then places disk $m$ along the generative spiral (giving $m$-number of parastichies along the direction 0-$m$), and also the $n$th disc (for the $n$-parastichies turning in the other direction).

Here he uncovers a geometric relationship, when he specifies that disks 0, $m$, and $n$—all with the same diameter—are in contact and placed along the generative spiral with a constant divergence angle. In this way, he can link the disk diameter $b$ to the divergence angle. In other words, saying that the disks have a given diameter $b$, and that they are in contact, results in only one possible value for the divergence angle. **The Van Iterson diagram is the plot of this divergence angle as the disk diameter changes.** It turns out that this relationship, given the appropriate coordinates, draws a simple arc of a circle.[7] Each arc of the circle then corresponds to a family of lattices of parastichy numbers $(m, n)$ whose shape varies as the diameter $b$ decreases.

We do not, however, see the whole arc of the circle. We see *parts* of circles, which get smaller as we move downward on the diagram. When the next disk on the diagonal, $m + n$, comes in contact with disk 0, our first arc comes to an end (see figs. 10.3 and 10.4). Then two new solutions and two new arcs of circles become possible: one corresponding to a contact between disks 0, $m$, and $m + n$, and the other to a contact between disks 0, $n$, and $m + n$. Each of these two arcs of circles will also come to an end, giving rise to two new arcs, and so on. This constructs a wonderful self-similar tree.

## The Tree, in More Detail

Here, let's zero in on Van Iterson's tree, as enlarged in figure 10.4. Each point in the plane represents a lattice. The horizontal axis represents its divergence angle, and the vertical axis the difference in height between its disks 0 and 1.

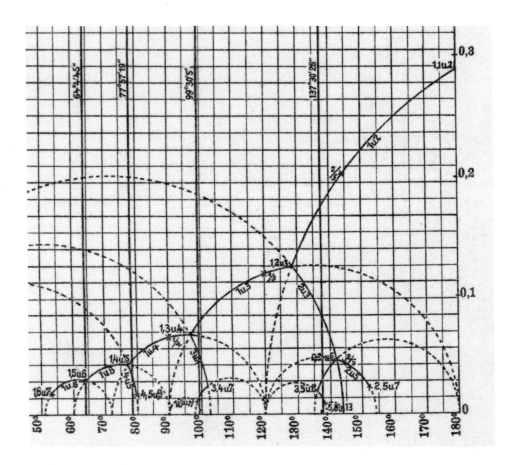

**FIG. 10.4** An enlarged view of the all-important Van Iterson diagram, published in 1907; this beautiful arching tree formed by solid lines lies at the heart of much phyllotaxis research. Do you see the special points where one dark arc of a circle splits into two new ones?

Understanding the tree's labeling requires a quick German lesson: *und* means "and." In this larger view of the tree, you can see that Van Iterson has labeled the topmost branch "1 u 2," with *u* being the abbreviation for *und*. At the branching point below, "1, 2 u 3" yields the two branches "1 u 3" and "2 u 3."

In general, two new arcs of circles can be seen at the end of each branch *m* and *n*: *m* and *m* + *n* in one direction, and *n* and *n* + *m* in the other. Continuing the computation, the tree emerges as a solid line. It starts from the simple (1, 1) opposite parastichies—whose endpoint appears at top right, at 180° and roughly 0.3. Next, Van Iterson maps (1, 2) phyllotaxis, following the first arc down to about 130°.[8] Continuing on down, he maps (1, 3) and (2, 3). The tree continues in this way, showing all the pairs of integers that are relatively prime.

The tree of possible solutions also has the remarkable property of being similar to itself, as Van Iterson noted. Bifurcations of branches are similar to one another—displaying the same angles and reflecting the similarity of the corresponding arrangements of disks.[9] It is disappointing that Van Iterson's observations here, which were far ahead of their time, went unnoticed by other scientists.

In figure 10.5, we have matched Van Iterson's disk lattices to points on the Fibonacci branch of his tree diagram. For every arc, the endpoint corresponds to a lattice with dense hexagonal packing, as shown in the disks at right.[10] Between these endpoints, the lattice expands and passes through a square state, as seen in the lattices at left. In the square state, there is maximum space between primordia. As illustrated in figure 10.5, Van Iterson envisioned the system going from one square state to the next, with minimal change of angle. This approach selects the next "good" patterns—those along the Fibonacci branch.[11]

As we will see, Van Iterson proposed that this special transition from square state to square state is actually what happens during a plant's development. Starting from the first possible arrangement of (1, 1), the plant then selects the sequence of Fibonacci numbers, with the divergence angle between leaves converging toward the golden angle of approximately 137.5°. In addition, Van Iterson offered a satisfying resolution to the divergence angle debate, his own answer falling between the Schimper-Braun rational divergence angle assumption and the Bravais brothers' golden angle one. According to Van Iterson, neither is right, but both are close. In his own view, the divergence angle zigzags between the rational lattices of Schimper-Braun and the golden angle ones of the Bravais brothers, getting closer to the latter as the plant develops.

Apart from this extensive theoretical work, Van Iterson also ran some simple practical experiments, looking at disk stacking using disks of different sizes.

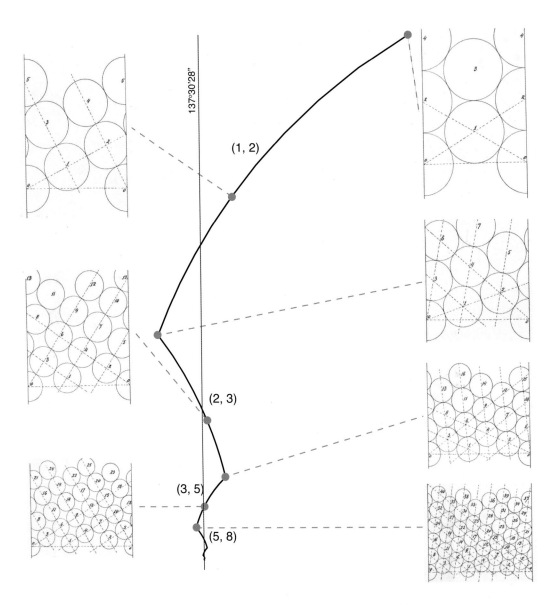

**FIG. 10.5** This figure shows how square lattices (at left) and hexagonal lattices (at right) correspond to points on the Fibonacci branch of the Van Iterson diagram. Do you see the difference between the left-hand patterns and the right-hand ones? Are they placed differently on the arcs?

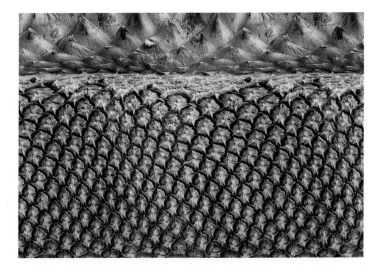

**FIG. 10.6** This unrolled pineapple offers a beautiful stroll through the branches of Van Iterson's tree. In the lower part, the scales form a square lattice, each scale sharing edges with four neighbors. In the upper part, the packing becomes hexagonal and scales share edges with six neighbors.

## Models of Lard and Wax

In search of a physical model, Van Iterson got creative. He wanted a way to place models of tangent circles of varying sizes on the surface of a cylinder. His idea? He placed a wooden pole inside a glass cylinder, using a ring at the bottom to hold the pole in place. He could then sandwich wax disks between them in order to observe the transitions between different patterns.

He was quite proud of his recipe for disks—which must have smelled pretty bad—based on a wax used by engravers to transfer patterns from one surface to another:

½ kilo pitch
1 kilo beeswax
50 grams Venetian turpentine
Lard, as needed
Flour, for sprinkling disks

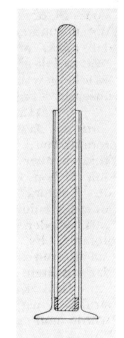

**FIG. 10.7** Van Iterson's wooden pole inside a glass cylinder, used to model disk packing.

After melting together the first three ingredients, he then kneaded the mixture with enough lard to obtain "the desired plasticity."[12] He then sprinkled the disks with flour to prevent them from sticking together once wedged between the glass cylinder and the wooden pole. He felt this concoction was far superior to any other. "The material remains stable even after years," he wrote, "and the old disks can immediately be reshaped."[13]

Van Iterson recognized the limitations of his model. It required "quite a bit of imagination,"[14] he wrote, to make the leap from circles on a cylinder to the forms actually found in nature. He knew it would be more accurate to use complex forms, such as a conical surface with elongated leaf shapes he called "folioid." He noted, however, that "the making of such structures is quite time-consuming and cannot be carried out as simply as with circular disks along a cylindrical surface."[15]

**FIG. 10.8** Another type of physical model explored by Van Iterson, using wooden balls of uniform size. Do you recognize the phyllotaxis patterns?

## Spiraling into the Future: Renormalization

With his extensive computations, close to pure mathematical work, Van Iterson introduced a fundamental diagram that is an early example of a self-similar structure. (To read Chris and Stéphane's debate over whether the diagram is actually fractal, see chapter 15.) It is unfortunate that the originality of his work remained unnoticed until recently. One factor is undoubtedly that although Van Iterson tried to address his work to plant scientists, the content was heavily mathematical.[16] Moreover, his later research strayed far from phyllotaxis, concerning mainly applied botany—from improved rubber production to the microbes used by the Royal Dutch Yeast and Spirits Factory.

Not only did his work show the self-similarity of the tree of disk lattices, but Van Iterson also used the tree to show the quantitative, geometric relationships between the transitions at different disk sizes and parastichy numbers. His approach, later known as "renormalization," would in a totally different context earn physicist Kenneth G. Wilson a 1982 Nobel Prize.

While renormalization techniques have been applied in many physical and mathematical settings, they all share a common thread. They compare the relationships between a phenomenon seen at one scale using one parameter value against the same phenomenon at another scale using a different parameter value. For Van Iterson, the phenomenon was the transitions of parastichy numbers in the context of disk packing. The parameter he used, $b$, was the size of disks relative to the size of the cylinder—that is, the diameter of the young plant organ relative to the circumference of the stem. He then drew the transition patterns for different pairs of parastichy numbers, at different values of $b$, and noticed that the transition patterns repeat at different scales.

Like Schwendener before him, Van Iterson modeled phyllotaxis transitions by using disks of decreasing radius stacked one by one. The decrease reflects the fact that as the stem grows fatter, the primordia are relatively smaller. But the real beauty of Van Iterson's view is how clearly he shows geometric changes in the pattern as the plant grows, relating these to his tree diagram. In figure 10.9, the disk diameters are largest at the left and become progressively smaller,

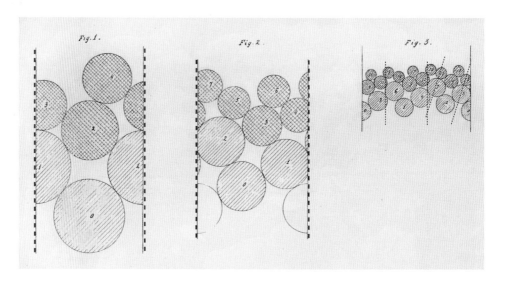

**FIG. 10.9** Here, Van Iterson has drawn primordium disks of decreasing radius as the plant grows. Can you see the same first two patterns in the more complex third one? This type of rescaling will later be called renormalization.

representing the maturation of a plant. At left, the figures represent transitions from (1, 1) to (1, 2) phyllotaxis; in the center, from (1, 2) to (2, 3) phyllotaxis; and at right, from (3, 5) to (5, 8).

Now the beautiful connection to Van Iterson's tree becomes clear. Using dashed red and blue lines, we have marked how rescaled and tilted copies of the disks in the two left-hand figures sit inside the figure at right. And guess what? The rescaling factor is none other than the powers of the golden angle. In doing this calculation, Van Iterson assumed that the top and bottom layers consist of disks zigzagging at right angles, like a row of disks in a square lattice. (Note that zigzags will reappear in chapter 14, in the discussion of fronts.) He then plugged in the value of $b$ for square lattices on his tree to arrive at his golden angle conclusion.

Van Iterson's renormalization technique was later independently rediscovered by Wilson in the early 1970s while he was working on the problem of phase transition—how liquid water becomes water vapor, for example, and how, at a

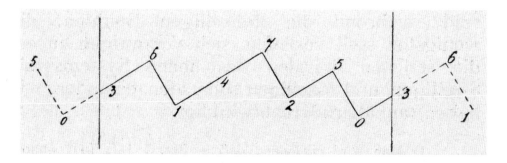

**FIG. 10.10** To make his argument a little tighter, here Van Iterson focuses on what he calls the *zickzacklinie* (zigzag line), representing the highest layer of disks at a given time. Similar zigzags will appear later in this book. Do you recognize from Hofmeister's drawings the last row of primordia before the meristem?

critical value of temperature and pressure, it becomes both liquid and vapor. A powerful tool, renormalization is still intensively used in many physics theories and phenomena, as well as in purely mathematical contexts. From writing these relationships at different scales, universal constants emerge. As we have seen, in the case of Van Iterson, that universal constant is the golden mean.

In the field of phyllotaxis, Van Iterson's main achievement lay in seeing underlying mathematical structures. He found the same branching structure that Schimper and Braun used in their fraction trees, but now firmly rooted in disk lattices. In addition, Van Iterson did important work identifying the disk patterns that could explain transitions between different spirals on a single plant, and in particular between Fibonacci spirals. Even if his work on transitions was lacking in some respects, he saw them as essential to solving the phyllotaxis puzzle.[17]

As for the Van Iterson diagram, it has proven vital to seeing that the solution to spiral phyllotaxis is not fixed in advance but instead has a range of possibilities. Introducing a constraint—primordia in compact disk stacking—had a deep impact on the selection of possibilities. But alone, his diagram could not solve the puzzle of Fibonacci numbers in plants: the Fibonacci branch is only one among infinitely many branches. We still need a clear constraint to explain why plants select the Fibonacci one.

In the final analysis, Van Iterson's work was limited by the technology of his era. "We still have boxes of his analyses and calculations on large sheets of graph paper!" noted an archivist at Delft University of Technology. "When his widow visited us, she wondered how much more he might have done if he'd had a computer."[18] For that enormous transformation, we'll need the protagonist of the next chapter, Alan Turing.

## Coda: The Botanical-Mathematical Artwork of Arthur Harry Church

Van Iterson was inspired by the phyllotaxis investigations of Arthur Harry Church, who believed that all plant patterns could be explained through mathematical calculations. Ultimately, however, Church's artwork has generally proven more enduring than his botanical studies.[19]

For Stéphane, Church's work reveals a touching contradiction between a too-perfect idealization of nature and the imperfections of reality. In the drawing in figure 10.11, for example, Church depicts some realistic details in the central petals, including hairs, veins, and wrinkles. Yet overall the flower is drawn as perfectly symmetric, with smooth curves, especially in its upper petals and stamens. Real flowers and stamens would display more "accidents of life."

Church's tendency toward imagined perfection is even more visible in the flower "diagram" at lower left. It shows a stamen arrangement that follows regular Fibonacci phyllotaxis, with 13 nearly radial spirals. Similarly, the plant spike at right has very regular architecture, showing the self-similarity of the plant. (Incidentally, this idealized form can be nicely described using the L-systems we will see in chapter 12.) Church had such an idealized view that he described phyllotaxis as giving "perfect mathematical expression"[20] to the spirals on pinecones and sunflowers. In his opinion, this pattern is "so perfect that any deviations from it in the actual plant must be due to the influence of some extraneous force not yet considered."[21]

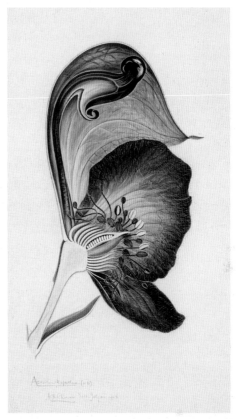

**FIG. 10.11** Church's drawing of seed formation in a monkshood (*Aconitum napellus*).

**FIG. 10.12** In this beautiful watercolor, Church has painted the same view of the somewhat idealized monkshood shown in figure 10.11.

PART IV

# Have Computers Shed Any Light?

# CHAPTER 11

# Sunflowers on Turing's Primitive Computer

Did
he
find truth
in daisies?
For Turing, simple
steps led him to complexity,
but not to happiness. He died unlauded, alone.

Phyllotaxis ran like a red thread through the short life of mathematician Alan Turing (1912–1954), from his childhood in England straight through to his heartbreaking death. In his final years, his phyllotaxis investigations marked a crucial turning point—Turing being the first to use a computer to tackle the question of why Fibonacci numbers appear in plants. His work on this problem was unfinished and for a long time unpublished. But in recent years scientists have pieced together the advances made by this iconic genius, whose ideas were essential to developing the modern computer.

When he was about 10, Alan first discovered natural science in an American children's book he received as a gift, according to Andrew Hodges's classic biography *Alan Turing: The Enigma*. Using simple language to describe developmental biology, *Natural Wonders Every Child Should Know* delved into such

**FIG. 11.1** Turing was known for his irreverent sense of humor.

**FIG. 11.2** Turing's mother, Ethel Stoney Turing, drew this picture of Alan getting distracted during a game of field hockey. Perhaps he was already thinking about plant spirals?

topics as the growth of starfish eggs and the inner workings of the human body. From the moment Alan read these pages, his brain was off and running.

Although he struggled at first to find intellectual focus, he ultimately obtained a degree in math from Cambridge University. Then came World War II. Britain needed his extraordinary mind if the country was going to get through the war. At Bletchley Park, Turing led the effort to crack the Nazis' secret codes, designing the electromechanical rotor machines that the cryptologists called "bombes." (This code-breaker chapter of his life was dramatized in the 2014 movie *The Imitation Game*, with Turing played by Benedict Cumberbatch.)

At Bletchley, daisies would make an appearance. Turing had become friendly with Joan Clarke, a high-level code breaker. She was the only female employed to practice Banburismus, the difficult cryptanalytic process that Turing invented to help crack the German navy's secret messages, sent by their Enigma machines. (Doing Banburismus required a mastery of sequential conditional probability, used to make educated guesses about each day's coding in order to narrow down the decryption possibilities.)

It was intense, high-pressure work with many lives at stake: their decodings helped stop German submarines from sinking British and American ships. Yet Alan and Joan did take occasional breaks. One afternoon, as Hodges described it, while lying on the lawn of Bletchley Park they started looking at daisies. Joan explained how to find the generative spiral, counting the number of leaves, then counting the number of turns between one leaf and the next that grows almost directly above it. Often, these are Fibonacci numbers. "They made a series of diagrams to test this hypothesis which did not satisfy Alan," Hodges wrote, and he "continued to think about 'watching the daisies grow.'"[1]

Did Clarke pick a flower and pull off the petals in a game of "he loves me, he loves me not"? Probably not. Yet as the friendship developed, Turing saw the possibility of a companionship that might work. After confessing that he was attracted to men—Clarke was undeterred—he proposed marriage and brought her home to meet his family. In the end, however, the relationship fizzled.

## Computing Biological Patterns

For all his brilliance, Turing was down to earth and interested in the practical side of things. In his early work on computable numbers, he didn't start from abstractions but instead from the real people employed to do calculations—the early "computers," who were often female. Based on their basic approach to doing calculations, he envisioned a theoretical machine that, given enough time, could compute anything desired and even reproduce human thought.[2] Turing's practical electronic work during the war led him to design an actual machine that would work like his theoretical one. But once the war was over, his design went nowhere, and the computer was more successfully developed by others.

Not until several years after the war did Turing turn in a serious way to the riddles of biological patterns.[3] He had a new tool at his disposal: the world's first commercial computer, a Ferranti Mark 1. And he had, as usual, a radically new way of looking at the world. In February 1951, Turing wrote a letter to biologist John Zachary Young saying that he was working on a mathematical theory that he believed would explain many patterns in nature, from the spots on animals to Fibonacci numbers in plants.

Just a few months later, Turing completed his landmark paper "The Chemical Basis of Morphogenesis." Decades ahead of its time, the paper would turn out to be seminal to the fields of morphogenesis and biomathematics. In these pages, published by the British Royal Society, Turing laid out his theory of morphogenesis through chemical reaction and diffusion.[4]

> At present I am not working on the problem at all, but on my mathematical theory of embryology, which I think I described to you at one time. This is yielding to treatment, and it will so far as I can see, give satisfactory explanations of -
> i) Gastrulation.
> ii) Polyogonally symmetrical structures, e.g., starfish, flowers.
> iii) Leaf arrangement, in particular the way the Fibonacci series (0, 1, 1 = 2, 3, 5, 8, 13,.......) comes to be involved.
> iv) Colour patterns on animals, e.g., stripes, spots and dappling.

**FIG. 11.3** In this letter to zoologist John Zachary Young, Turing wrote of his quest to explain the Fibonacci series in plants. Could understanding phyllotaxis lead to a more general understanding of pattern formation in nature?

As a starting point, he tried to understand how primordia in plants might form in the ring around the meristem (see his red ring drawing in fig. 11.6). He then approached the problem using chemistry, asking how a seemingly homogeneous chemical mixture could be unstable, creating spots where one molecule is more concentrated. To answer this question, he proposed a theoretical model: Two chemicals are homogeneously spread on a surface. One is called the activator, and the other is called the inhibitor. The two begin uniformly mixed, in an unstable equilibrium. When there is a small fluctuation, the amount of activator can suddenly rise. But this rise also increases production of the inhibitor, leading to simple oscillations of the two chemical concentrations.

Here is where Turing's brilliant insight comes in: he takes this basic scenario and then examines how these two chemicals diffuse *in space*. The rise of the inhibitor is reduced by lateral diffusion, first leading to a stationary peak of the activator. This spreading inhibits the growth of another peak nearby. But it can also favor the growth of another activator peak farther away. In other words, where there was once uniformity, now a stable pattern emerges.

As shown in figure 11.4, Turing used data generated by his Ferranti Mark 1 computer to understand how chemical reactions might lead to spots. Turing's shadings and blobs indicate places where the reactant values are higher. The 32 teleprinter symbols in the background indicate reactant values. Since the computer had a very primitive screen, he had to painstakingly copy out the list of printed-out results by hand, placing each value in his simulated space. He drew two wavy outlines to indicate the "river" of higher values, with highest activator values shaded dark gray. And voilà: spots.

The idea is very similar in phyllotaxis and animal spots, Stéphane explains. "You start with the simplest step, an instability creating

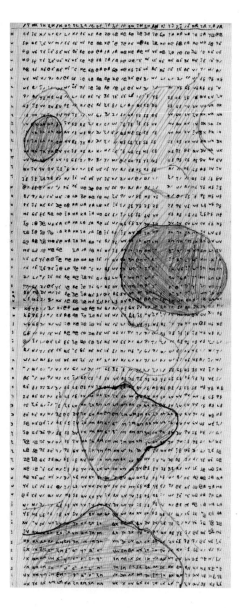

**FIG. 11.4** Using the world's first commercial computer, Turing mapped—by hand—the results of chemical reactions trying to reproduce the successive appearance of primordia. But doesn't this look more like animal spots?

a local concentration bump—a new primordium on the growing plant. Then you repeat the process. There is no reason why this cannot lead directly to any complex result, such as Fibonacci spirals in plants."[5]

Not until some 30 years later would the real significance of Turing's work on chemical morphogenesis be felt. This was when theories of chaos and dynamical systems began to explain instabilities—how a system can spontaneously change its state and its symmetries. Turing's paper then became very influential. He was the first to show that a pattern could arise spontaneously from an instability, without any internal prepatterning or external influence. In another nod to his originality, the concept is now known as "Turing instability." This instability helps explain many patterns in nature, including the plant patterns that would draw Turing's attention in his last painful years.

**Sunflowers and a Hypothesis**

After completing his groundbreaking morphogenesis paper, Turing had a fling. At the time, as he would soon discover, this act was exceedingly dangerous. When Turing rashly reported to the police a minor theft—committed by an acquaintance of his lover—the investigation took off in a very different direction. Turing was charged with "gross indecency," a term used at the time to prosecute homosexuals. Submitting to a particularly cruel and humiliating practice, Turing consented to take estrogen for a year in order to stay out of prison. The chemical hormones made him impotent, and he began developing breasts. Although he continued to do independent research, Turing would never publish a scientific paper again.

Still, he continued his investigations as best he could. In May 1953, just a year before his death, Turing wrote a letter to mathematician Donald Coxeter, an expert in geometry, about his ongoing phyllotaxis project:

> According to the theory I am working on now there is a continuous advance from one pair of parastichy numbers to another, during the growth of a single plant. . . . You will be inclined to ask how one can move continuously

from one integer to another. The reason is this—on any specimen there are different ways in which the parastichy numbers can be reckoned; some are more natural than others.

During the growth of a plant, the various parastichy numbers come into prominence at different stages. One can also observe the phenomenon in space (instead of in time) on a sunflower. It is natural to count the outermost florets as say 21+34, but the inner ones might be counted as 8+13. . . . I don't know any really satisfactory account, though I hope to get one myself in about a year's time.[6]

Turing did not live to publish his work on plant patterns, and for many years his notes and pen-and-ink drawings remained uncollected and unexplained. Then in the early 2000s, British mathematician Jonathan Swinton dove into the materials. His efforts culminated in an illuminating 2013 paper, "Turing, Morphogenesis, and Fibonacci Phyllotaxis: Life in Pictures." As Swinton notes,

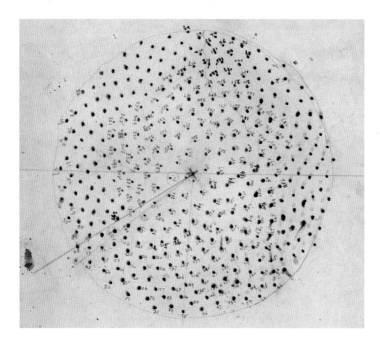

**FIG. 11.5** Here, Turing has numbered the florets on a sunflower, with 55 and 34 spirals toward the outer edge, 21 and 34 spirals closer to the center, and so on. Wait, is this plant actually a *Carlina* thistle? See the note on mistaken identity below.

Turing's drawings tell the story of his "quest to explain, among much else, the appearance of Fibonacci numbers in the natural world."[7] And like so many of Turing's theories, the concepts in the drawings were far ahead of their time.

His red ring made for an especially powerful visual. In Turing's 1952 article on morphogenesis, he had reduced everything to a ring where the essential chemical reaction occurred to create spots. A year later, while still trying to solve the puzzle of phyllotaxis, Turing expanded his original model to include the lateral development of spotlike structures in plants—the primordia. In the drawing of bud growth in figure 11.6, Turing depicts the meristem tip as an empty circle at the center. Surrounding this zone is the red ring where the primordia first appear. As the stem grows, the primordia drift outward and increase in size, as shown by the larger black spots.

In his book *Alan Turing's Manchester*, Swinton described Turing's phyllotaxis work this way:

> Turing had created a mechanism for making spots appear, and the spots were separated by a typical length-scale, which depended in a complicated but tractable way on the numbers put into the model. The number of spots one might see developing in a narrow ring of tissue, as he used in his 1952 paper, would depend not only on that length-scale but the circumference of the ring.
>
> But now suppose the ring is no longer narrow, so that patterns can develop across as well as along it, in a way where the placement of new spots depends on where the older ones went. What Turing saw was that this dynamic model could allow the pattern to get more and more complicated as the inter-spot distance got smaller.[8]

In this way, just as we saw earlier with disk packing, complex patterns can emerge one circle at a time. Seeing Turing's work on this dynamical model led Stéphane to reflect on how it came to be. "I often found myself asking how Turing moved from his computing machine to the problem of phyllotaxis," Stéphane says. "What was the connection? Finally, I realized that they involved exactly the

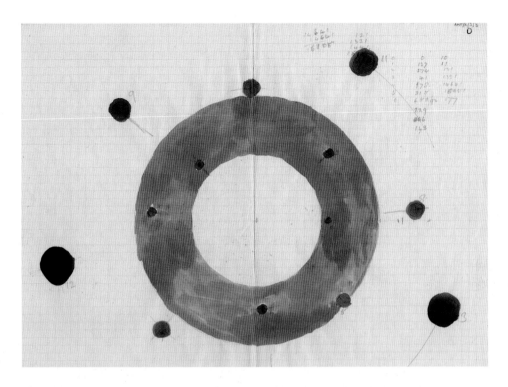

**FIG. 11.6** Here, Turing has drawn a model of the dynamic growth process in a bud. The red ring is where new primordia appear, depicted as small black spots. Can you see his numbering of the spots and his list of their angles?

same approach. With the computer, he looked at how a human makes computations, identifying the basic steps. He then reproduced these steps in the simplest possible machine, showing that it could do all the possible computations.[9] The main result is that a simple process, repeated over time, can lead to the most complex things you can imagine. This is also why Turing pioneered the idea of artificial intelligence."

His life cut short, Turing never completed his work of trying to understand *why* phyllotaxis patterns might occur. Seeking to solve the puzzle, he focused on the transitions between phyllotaxis spiral modes. Turing saw that parastichy numbers can change in only two possible ways—one way being from one pair of consecutive Fibonacci numbers to the next.[10] He then developed a hypothesis

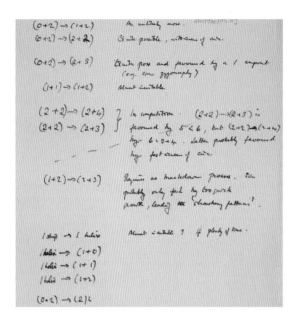 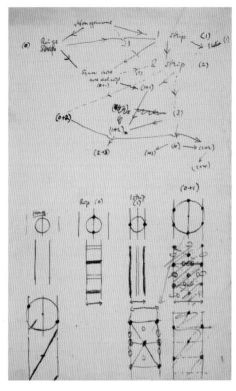

**FIG. 11.7** At left, Turing commented on possible transitions between the first parastichy pairs: "An unlikely move," "Quite possible," and "Almost inevitable?" At right, he diagrammed possible transitions between phyllotaxis modes, thinking aloud as he tried to solve the problem.

that there is a fundamental reason why plants select one solution and not the other. This theory, which he tested on his computer, was what Turing called his hypothesis of geometric phyllotaxis. Although he got most of the way to proving it, he died before he could finish.

"He was right," Swinton concluded. Turing's hypothesis for pattern creation "does hold for a wide family of mathematical models of spot formation in growing tissue."[11] Many years later, the hypothesis of geometric phyllotaxis was proven by Stéphane and his physicist colleague Yves Couder in Paris, who at the time did not even know about Turing's work on the subject. (For more on them, see chapter 13.) Not until 1992—nearly 40 years after they were written—would Turing's notes on his hypothesis of geometric phyllotaxis be published.[12] Coincidentally, Stéphane and Yves's solution was published at nearly the same time.

## Death

Theories about Turing's death from cyanide poisoning at age 41 have proliferated in recent years as his fame has grown. Was it suicide? An accident? Even murder? Turing's home laboratory in his house outside Manchester had equipment for electroplating, a process that used cyanide. When his housekeeper found Turing's body, a half-eaten apple sat by his bedside.

- Possibly, he ate the apple to cover up the taste of the cyanide he swallowed.
- Possibly, he chose to die by poisoned apple in a nod to his favorite fairy tale, "Snow White."
- Possibly, he wanted to make his suicide look like an accident to protect his mother's feelings, as his biographer Hodges has suggested.
- Possibly (though not likely), his death truly was an accident, from inhaling cyanide gas that leaked from his electroplating equipment.
- Possibly, his suicide was an *acte manqué*, using the cyanide gas he knew was dangerous in an accidental "failed act" that shielded his own consciousness from his desire to die.
- Possibly (though again, not likely), Turing was murdered.

In 1954, as the Cold War heated up, British intelligence might have considered him a security risk, as gay people were generally viewed as being vulnerable to blackmail, and Turing knew many wartime secrets. Then again, Turing's homosexuality had already been announced in several newspapers that covered his criminal prosecution, so how could he have been blackmailed? One thing is certain: Turing left no suicide note. Perhaps he would have savored the idea that even today, his death remains enigmatic.

Because Turing's phyllotaxis investigations remained unknown for decades, the next computer scientists to approach the Fibonacci puzzle had to start from scratch. But interestingly, Aristid Lindenmayer in the Netherlands would use a similar approach to Turing's, leading to computer-generated sunflowers and beyond.

## A Case of Mistaken Identity!

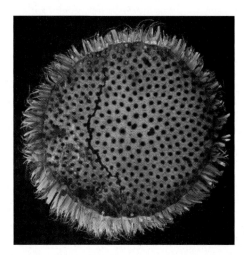

**FIG. 11.8** Although this photo of plant spirals from Turing's files is always identified as a sunflower, Stéphane knows from the French countryside that the flower is actually a carline thistle (*Carlina acanthifolia*). Turing must have taken the photo while on vacation.

**FIG. 11.9** A carline thistle photographed by Stéphane —definitely not a sunflower.

## Try Your Hand

### Count Spirals on Turing's Sunflowers

Turing saw that Fibonacci patterns often occurred in sunflowers, but he died before he had a chance to examine enough real plants to provide good data. To remedy this problem, a science museum in Manchester, England, sponsored a citizen science project in 2012, on the 100th anniversary of his birth. Participants grew sunflowers and counted their spirals. All told, data were collected from 657 seed heads.

Turing scholar Jonathan Swinton then led a team that analyzed the results and published a paper. They discovered that many sunflowers did not conform precisely to the Fibonacci rule—with the most common exception being seed heads with one spiral fewer than a Fibonacci number. "An unexpected further result of this study," wrote the authors, "was the existence of quasi-regular heads, in which no parastichy number could be definitively assigned."[13]

Although the project is over, its excellent worksheets might inspire you to grow your own sunflowers. Will you get Fibonacci numbers of spirals, or something unpredictable?

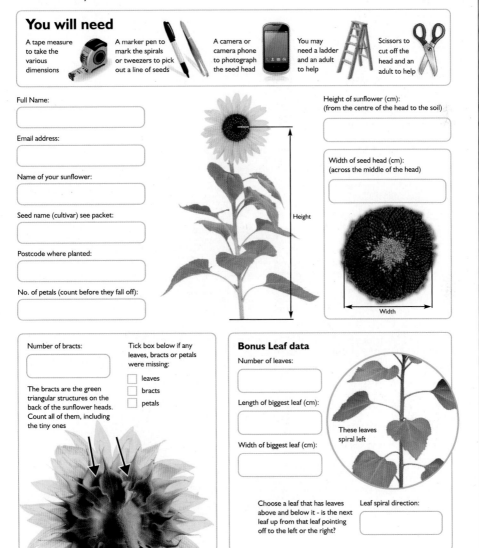

FIG. 11.10 A worksheet for recording observations of a sunflower.

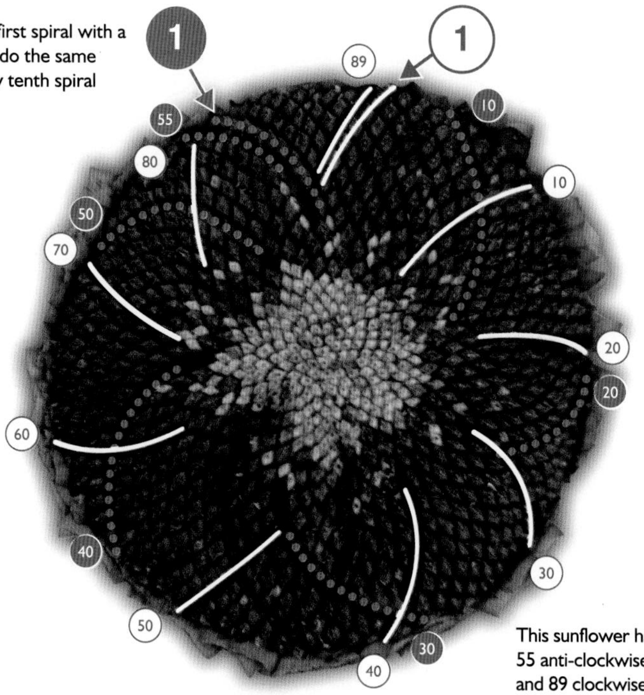

**FIG. 11.11** The worksheet's second page gives clear instructions for counting the two sets of spirals on a sunflower head.

# How to count the spirals

After you have measured and recorded the height of your sunflowers and the spiral patterns are showing clearly in the seed head:

1. Use scissors to cut off the seed head about 5 cm down from the top of the stem (get an adult to help with a ladder if needed).
2. Have a look at the picture below to help you identify the spiral patters that bend in two directions:

    ⟲ clockwise spirals   ⟳ anti-clockwise spirals

3. Use a camera to take a good picture with the spiral patterns clearly in focus. You will need to upload this later for us to check your spiral count.
4. Either print out the picture to mark on the spiral patterns with a pen or use tweezers to pick out a line of seeds from every tenth spiral (see our online video at: http://www.youtube.com/watch?v=A02ccEK7Hjw to show you how to and upload your photo for our experts to check). We have done both in our diagram below.
5. Count the spirals in both directions - did you get Fibonacci numbers (0, 1, 1, 2, 3, 5, , 8, 13, 21, 34, 55, 89, 144) Add them together to get a total, is this a Fibonacci number too?
6. If you didn't get a Fibonacci number, don't worry it's still important for us to know what you got so record your answers on the sheet and upload your photo for our experts to check.

**Don't worry if you can't tell us the direction of the spirals. The numbers are more important.**

Number of clockwise spirals:

Number of anti-clockwise spirals:

Total Number of spirals:

Mark the first spiral with a pen, then do the same with every tenth spiral

This sunflower has 55 anti-clockwise spirals and 89 clockwise spirals

CHAPTER 12

# Leaves and Petals as Data Points

A
plant
can be
as abstract
as you want it to
be. To understand a plant's growth,
just use L-systems or theoretical physics!

Starting in the 1990s, physicists returned to the field of phyllotaxis, following in the footsteps of crystallographer Auguste Bravais. The analogies they drew between physical models and plant growth had a profound impact on scientists' understanding of the Fibonacci riddle. They worked in new ways, sometimes experimental and sometimes purely theoretical. They ran computer simulations—Turing had already shown the way.

As they pushed their dynamical approach a few steps forward, new ways of explaining plant patterns emerged. As Ian Stewart wrote in *Scientific American*, "Genes need not tell [plants] how to space the primordia. That's done by dynamics. It's a partnership of physics and genetics."[1] Among the main advances were the following:

- Creating a new, abstract language for describing plant structure and growth (L-systems).
- Shedding light on the global efficiency of disk stacking.
- Letting the pattern grow one primordium at a time, according to simple rules (following Hofmeister), in both experiments and computer simulations.

Call it the Modern Paradigm.

**Early Computer Simulations**

In 1969, the theoretical biologist Aristid Lindenmayer (1925–1989) and his student Arthur Veen began using computer simulations to study phyllotaxis. Like many scientists, Lindenmayer led an international life: born in Hungary, he got his PhD at the University of Michigan and then taught at Utrecht University in the Netherlands. Although they had only limited computing power, he and Veen chalked up some successes and opened up the field to further attempts.

At the time of these early computer simulations, Lindenmayer was just coming off a major discovery. In the previous year, 1968, he had published a paper introducing what are now known as Lindenmayer systems (L-systems) to model plant growth. L-systems employ an ingenious type of computer language, using symbolic alphabets and formal grammar. These symbols represent different elements of biological systems at different scales as a way to show their evolution and growth.[2]

The computer that Lindenmayer and Veen used was only slightly better than Turing's. Still, they were able to model cells on a grid, showing how they interacted with neighboring cells according to simple rules.[3] Although they did not cite Turing, they used a similar model: a growth inhibitor diffuses and decays on a cylindrical stem that has been divided into square cells. Where the inhibitor level drops sufficiently, a new primordium forms.[4]

Since their computing power was very limited, they concentrated on showing that—with the right parameters controlling rates of diffusion and decay—they could reproduce patterns with a constant divergence angle. They used precise knowledge of the Van Iterson diagram to set up the parameter values and initial configuration of their simulations. But because their results were hamstrung by the low resolution of their simulations, they were not able to reproduce Fibonacci transitions.[5]

This grid looks very similar to Turing's handwritten computation of diffusion, with its shaded spots, that we saw in the previous chapter—only now the spots appear as green rectangles. In this version, a "stem" is built from horizontal rows of 26 characters, or cells. Each concentration of growth inhibitor is assigned a value, starting with numbers 1 through 9. For concentrations higher than 9, the symbols switch to letters, starting with A.[6] The inhibitor diffuses and decays over time.

Where the concentration of the inhibitor falls low enough, a "leaf" appears as a green rectangle.[7] At this point, the leaf itself becomes a source of inhibitor, as shown by the higher concentrations—letters—in surrounding cells. This means that a new leaf cannot appear close to an existing leaf. After a fixed number of time steps, a new layer of cells is added at the top of the cylinder as the plant grows taller.[8]

The work of Lindenmayer and Veen bore hallmarks of computer models that would be developed later, using point-like primordia and a simplified pattern formation process that evolves dynamically.[9] By fine-tuning the model's parameters and its initial configuration, later scientists sought to generate self-propagating patterns (or steady states) with a constant divergence angle.

```
30  5555555555555555555555555555
29  3333444443333333333333333333
28  2233344433322222222222222222
27  2223455543322222222222222222
26  2234689864322211112223333222
25  22358FHG8532222112233345443
24  33359H█!9543332222235798664
23  334586G1G865554332234 8FHF85
22  44457998767898643335 9H█H95
21  55666666568GHF 85334586 1G86
20  7899755456 9I█I9544457 99876
19  9G1G9644469G1G8655666666656
18  A1█1A6545579087678997 55456
17  9G1G9656666666569G1G964446
16  79A97679A9755456A1█1A65455
15  6666569G1G9644469G1G965666
14  755456A1█1A65455 79A97679A9
13  9644469G1G9656666666569G1G
12  A6545579A97679A9755456A1█1
11  9656666666569G1G9644469G1G
10  7679A9755456A1█1A6545579A9
9   569G1G9644469G1G9656666666
8   56A1█1A6545579987679A97554
7   468G1G9656666666569G1G9644
6   55799876 79997544469 1█19654
5   666666569G1G8543458G1G8655
4   98654446 91█19543346 8987668
3   GF7544458G1G85433 44565557F
2   █E75444568 9865444444 55557E
1   55555555555555555555555555555
```

FIG. 12.1 This early computer simulation models the vertical growth of a plant, with a green rectangle for each leaf. How many parastichies are there? Can you spot the island of 1's where a new primordium will appear?

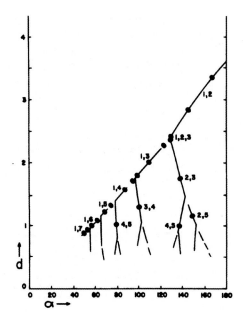

**FIG. 12.2** In their simulations, Lindenmayer and Veen showed that they could reproduce a few lattices on the all-important Van Iterson diagram, marking their successes with black dots. Why do you think some branches are broken?

```
#define S         /* seed shape */
#define R         /* ray floret shape */
#include M N O P  /* petal shapes */

ω  :  A(0)
p₁ :  A(n)  :  *  →  +(137.5)[f(n^0.5)C(n)]A(n+1)
p₂ :  C(n)  :  n <= 440                →  ~S
p₃ :  C(n)  :  440 < n & n <= 565      →  ~R
p₄ :  C(n)  :  565 < n & n <= 580      →  ~M
p₅ :  C(n)  :  580 < n & n <= 595      →  ~N
p₆ :  C(n)  :  595 < n & n <= 610      →  ~O
p₇ :  C(n)  :  610 < n                 →  ~P
```

**FIG. 12.3** Computer-generated sunflower from Lindenmayer and Prusinkiewicz's *Algorithmic Beauty of Plants*. Can you find the golden angle hidden in the programming instructions?

A generation later, in 1990, Lindenmayer coauthored *The Algorithmic Beauty of Plants*, the first book to describe the computer simulation of plant growth, including phyllotaxis, using L-systems. Computers had come a long way since Lindenmayer did his first phyllotaxis simulations. Together with Polish-born computer scientist Przemysław Prusinkiewicz, Lindenmayer created computer-generated zinnias, sunflowers, and roses.

They were able to add details of leaves and petals—and even shadows—according to the precise description of their positions. With these eye-popping images, the authors showed that Fibonacci numbers in nature can be expressed as elegant and fairly simple developmental algorithms, a set of rules that describe plant development over time. Their simulations were able, following Bravais, to directly encode parastichy number increases according to the Fibonacci rule.

## A First Physics Foray

In 1991, phyllotaxis entered a new phase when Leonid S. Levitov (1962–) drew a connection between a specific physical system and plant patterns. Levitov, a Russian-born professor of theoretical physics at MIT, did not do research on actual plants. For that matter, he didn't even do research on actual materials. As a theoretical physicist, he ran experiments using computer models of theoretical layered superconductors. From studying lattices of particles, he helped to prune Van Iterson's all-important tree, which had many branches that were not consecutive Fibonacci numbers.

What Levitov discovered and explained was this: phyllotaxis patterns appear in the flux lattice of a superconductor, at places where the particles minimize their repulsion energy within the system. By following the trail of these minima, *you travel only the part of the Van Iterson diagram that fits the Fibonacci rule.* Wow!

For those unfamiliar with layered superconductors, let's take a moment to explain. In his model, Levitov was interested in seeing how electricity flows through superconductors. More specifically, he wanted to see how the flow changed when he added more of the surrounding magnetic field. What he found was that the flow organization changed in ways similar to a plant's increasing parastichy numbers.

To visualize Levitov's work, it's helpful to know that when layered superconductors are exposed to a magnetic field, the electricity flows through vortices. ("Vortices" is the plural of "vortex," as in a tornado.) Here, "vortex" refers to the way flowing electricity generates a rotating magnetic field around it. These vortices then interact among themselves via the magnetic field they generate, repelling each other.

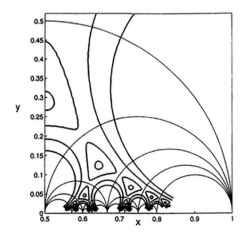

**FIG. 12.4** In Levitov's diagram, the small circles within triangles surround energy minima. These correspond to compact hexagonal packing in plants, a lattice of primordia with three parastichy numbers. Can you guess where the Van Iterson tree is hiding? (Hint: connect the minima.)

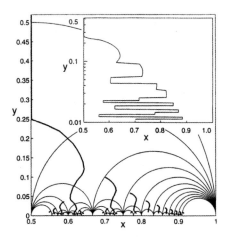

**FIG. 12.5** In Levitov's diagram, the boldface "crooked branches"—what remains of the Van Iterson tree —reveal crucial disconnections. How does this diagram match with the previous energy landscape? Can you guess what the thin half circles mean?

Levitov supposed that these vortices would organize into regular patterns, forming something like a familiar Bravais lattice of primordia (leaves) on a cylinder (stem).[10] When the surrounding magnetic field is increased, it reduces the effect of the interactions between the vortices. This allows them to pack together more tightly—similar to when Bravais decreased the rise of the generative spiral and packed the primordia into a compact hexagonal lattice. And voilà—in this way, the two models of phyllotaxis are comparable.

What Levitov found is this: by computing the best packing energy for all possible Bravais lattices, the best lattices—the ones that minimize energy, in the valley of this energy landscape—would lie on the all-important Van Iterson diagram. The reason is that when the vortices are nicely separated from each other by repulsion forces, they create equal space around them, similar to the distance between stacked disks.

But in doing his superconductor experiments, Levitov found an essential difference: the branches are actually *disconnected*. With this insight, Levitov advanced toward a proof of Turing's hypothesis of geometric phyllotaxis (see chapter 11), which states that there is a fundamental reason why plants select consecutive Fibonacci numbers when given two options. In Levitov's model, he talks about this choice occurring at a *bifurcation*. A bifurcation is a kind of physics fork in the road, where a small change in parameter values causes a sudden shift in behavior.

To better see this, let's dive into Levitov's diagram, figure 12.5, for a moment. Focus on the boldface "crooked branches" that rise up from the x-axis. Just as for the Van Iterson diagram, here each section of a branch corresponds to a lattice with parastichy numbers ($m$, $n$). Viewing figure 12.5 as an elevation map, you can see that each crooked curve starts from a dip, rises up, crosses a thin arc at a "mountain pass," and then curves downward again.

In addition, you can see that all the branches have a top ending: they stop precisely at a high-energy "mountain pass." If you compare this with Van Iterson's diagram, you see that there the branches do not stop. Instead, they connect to form a fused branch. These are the bifurcations. But in Levitov's diagram, there is only one path. Van Iterson could compute his solutions geometrically, but many were not valid *energetically*. Therefore a plant would not follow this path, according to the laws of physics.

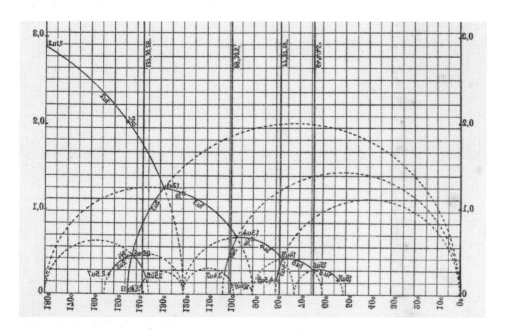

**FIG. 12.6** Van Iterson's tree again, but flipped, for comparison. Are the half circles the same?

This discovery has a major impact. With the bifurcations removed, we no longer have to guess which new solution to take. If we start from high enough in the diagram, we can follow only one continuous branch, the one that traces parastichies increasing according to the Fibonacci rule.

By using this property, Levitov was able to partly prove the hypothesis of geometric phyllotaxis (for regular Bravais lattices). To do so, he needed only to look at one transition around a minimum and use the beautiful symmetry apparent in his repulsion function—which also underlies the symmetries and self-similarity of the Van Iterson diagram. So, what was true for one bifurcation would be true for all the others.[11]

Still, Levitov's repulsion potential model left a major question unanswered. It assumed that the primordia form a lattice, before this lattice can be deformed into an optimal one. But how is the initial lattice formed in the first place? And what would cause this botanical deformation? Recall the first sentence that Simon Schwendener ever published: "All organic life is based on the game of continuous formation and deformation." In addition to proposing the theory of *deformation* by contact pressure, he also proposed a geometric mechanism of *formation*. It seems that the explanation for formation was missing from the work of Levitov.

Shortly after Levitov's work was published, two French physicists conducted an elegant physics experiment—far from the realm of plants—that shed light on the formation side. They figured out a way to generate Fibonacci spirals spontaneously, using magnetic droplets that start at the center of a disk and spin gracefully away. The video they posted of this Fibonacci dance has been viewed thousands of times. One of these physicists is someone you've already met.

CHAPTER 13

# The Big Experiment with Tiny Droplets

Can
you
conjure
spirals in
a lab? Well, just try
making magnetic droplets spin!
An elegant experiment, it caught the world's eye.

Stéphane's work in phyllotaxis began with a vegetable. Back in 1991, he was doing science research as an alternative to military service, while working as a private math tutor to make some money. On the way home from a fancy neighborhood in Paris, he passed a market that sold chic vegetables. There, he saw his first Romanesco broccoli, at a time when it was still a rarity. "What?!" he recalls thinking. "It's fractal!" He bought one and took it back to his lab, where he was spending 18 months under physicist Yves Couder investigating turbulence.

Stéphane showed the broccoli to Yves, who said, "You see these nice spirals? They seem to be exponential." But that was hardly enough to make Stéphane's heart pound. "I dismissed the idea," he said. "The fact that it was exponential wasn't very exciting. It simply showed that the broccoli was growing everywhere at a similar rate." Little did Stéphane suspect that this vegetable would take the

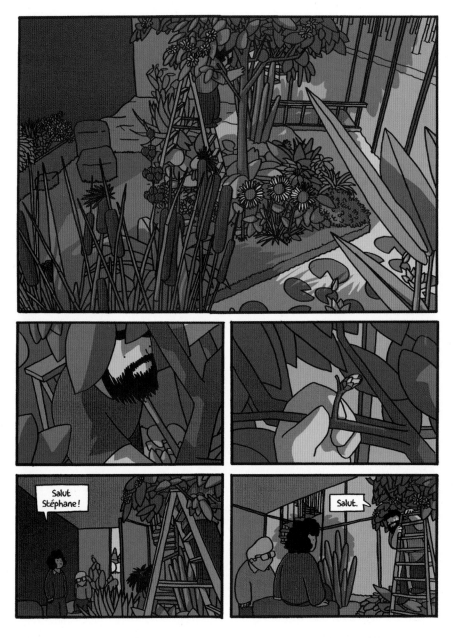

**FIG. 13.1** Stéphane and his plant lab appeared in a French graphic novel called *Le chercheur fantôme*, published in English in 2023 as *The Phantom Scientist*.

**FIG. 13.2** Romanesco broccoli, the vegetable that launched Stéphane on his phyllotaxis adventure. How many spirals do you see?

two of them on a journey beyond the work of Leonid Levitov, using dynamical physics to understand the Fibonacci problem better than anyone ever had before.

During his winter break, Stéphane traveled back to his family's country house in Corrèze, a beautiful part of southwest France with rolling hills, medieval villages, and humid forests with lichen-covered trees. In a quiet moment, he found himself idly gazing at the shelves of the ancestral library. Although Stéphane's great-grandfather Rémy Perrier had been an expert on insects, he'd left behind books on other branches of biology, including Philippe van Tieghem's 1884 introduction to botany.[1] "I thought I'd look at the book for fun, just to flip through the nice pictures," Stéphane said. "But then I found a drawing of a mathematical spiral. 'What's this doing here?' I wondered. 'And is it the same as what we saw

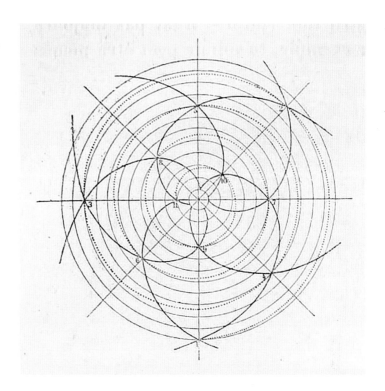

**FIG. 13.3** Leafing through a nineteenth-century book on botany, Stéphane stumbled upon this mathematical diagram. "What's this doing here?" he wondered. Were these spirals the same as the ones on his Romanesco broccoli?

on the broccoli?'" He read through the chapter and saw that the spirals were universal in plants and very often Fibonacci.

"I didn't believe it, of course," he said. "It was just an old botany book from the nineteenth century."

Now Stéphane was curious. "So I went out in the forest, and I picked up a pinecone from a *Picea*, a spruce tree," he recalled. "I counted the spirals on five or ten cones, and the spirals were always Fibonacci. I looked at a larch tree, and it was the same. I was really shocked, because I realized I'd never really looked at a pinecone before. My next thought was, if this is universal, as the book says, then it must be robust. If it's very robust, it must be simple. And if it's simple, I can reproduce it."

Back at the Statistical Physics Laboratory of ENS in Paris, Yves let Stéphane take some time off from turbulence (as they'd already given the military good results) to work on plants. Together, they designed a few experiments on plant spirals. Stéphane first tried out ideas that were very close to Schwendener's observations on contact pressure (see chapter 9). "If it's an asymmetric configuration and you push on it, then the pieces should slide in one direction, not the other," Stéphane said. "I tried to do this with balloons, inflating them to see if they'd twist. But I didn't get very far."

For inspiration, next they combed through other sources. Most helpful was a book by French botanist Lucien Plantefol that described the history of phyllotaxis in the introduction.[2] (In French, *plantefol* means "crazy plant"—quite a name for a botanist.) Reading that volume, Stéphane and Yves discovered the work of Hofmeister, whose idea was that plants grow step by step, maintaining distance between the primordia. "That," said Stéphane, "was really a great idea."

What they wanted to find was a simple model for packing primordia. "Physicists always love to simplify and approximate," Stéphane noted. They came up with the idea of stacking particles that repel each other. (This was getting close to Levitov's idea, although they didn't know about him yet.) As it happened, at the time, Yves was teaching a class with an expert in ferrofluids, liquids with tiny magnetic particles suspended in them. The light went on.

Yves told Stéphane that if they designed an experiment using magnetic fluid and a vertical magnetic field, then the droplets would repel each other—exactly the way new primordia distance themselves in plants.[3] The idea led to a demonstration of droplets performing a spiral dance that has become iconic in the field of phyllotaxis. The two physicists showed that the dynamics of plant growth can be reproduced with a physical experiment alone. Using repelling particles, their model offered a demonstration of dynamical packing.

To take a step back for a moment, let's briefly introduce dynamics, the branch of physics that investigates the motion of solid objects in time when affected by force. Its roots can be traced back to the experiments of Galileo Galilei in the late sixteenth century, who defined the law of motion for falling bodies by studying a small bronze ball rolling down an inclined plane. This view remained limited to mechanics. Three hundred years later, the field of dynamical systems emerged, describing the behavior of any system whose state changes over time. A key figure in the development of dynamical systems is French mathematician Henri Poincaré, who, at the turn of the twentieth century, published two classic papers on celestial mechanics, the "three body problem." This was a seed for the forthcoming theory of chaos.

## How to Make Magnetic Drops Spin

But let's get back to Yves and his magnetic fluids. "It was not an easy experiment," Stéphane recalled, "and at first, it didn't work." The setup involved the following:

- A round Teflon plate covered with silicone oil.
- A dome-shaped bump sticking out above the oil at the center of the plate, playing the role of the meristem apex.
- Magnets (Helmholtz coils) surrounding the plate.
- A pipette to release the magnetic droplets.
- A syringe plunger to push ferrofluid from the pipette.

To run the experiment, droplets of magnetic fluid were released from the pipette at regular intervals using the syringe plunger. The droplets were polarized by the surrounding magnetic field and drawn toward the edge of the plate. Falling on the central bump, each new droplet moved to where it felt the least repulsion of the other droplets, especially that of the closer and newer droplets. In other words, each droplet followed Hofmeister's rule for the placement of new primordia, finding the largest available space.

Fortunately, the droplets behaved. After first moving apart from each other, they simply traveled radially with the surrounding magnetic field toward the border (where they disappeared, falling into a "ditch"). The droplets' radial movements are analogous to what happens on a growing stem, where the primordia expand without changing their relative positions. (If this were not the case, the primordia would keep reorganizing, and their pattern would be constantly changing.)

  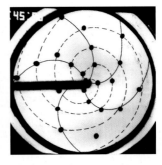

**FIG. 13.4** These photographs show three stages of the experiment, with the magnetic fluid droplets released faster each time. (You can see the pipette, the spinning droplets, and the outer ditch for "dead" ones.) Can you follow the motion of the droplets as they appear?

## The Droplets Do Their Fibonacci Dance

How do the dancers do their steps? When the magnetic droplets are released slowly, the new droplet simply moves in the opposite direction from the previous one. The other droplets are too far away to repel anything. But as the droplets are released faster, their dance of repulsion and radial motion leads them to organize in spirals. And these spirals display ascending Fibonacci numbers. Here, the patterns show parastichy numbers of (1, 1), (1, 2), and finally (5, 8). The spirals connecting the drops were hand drawn on the photograph by Yves.

The first page of Yves's lab notes in figure 13.5 shows the design of the experiment. One by one, droplets of magnetic fluid get released onto a Teflon plate with a central bump. The plate is covered with silicone oil ("Huile 47 V 100").

Surrounding the plate are Helmholtz coils that produce magnetic fields, with specific input currents ("$I = 3.5$ A"). As the drops are released, they float a bit in the oil, repelling one another while being attracted to the edge of the dish. (In this first version, there is no ditch.)

The first try doesn't work (off-center, etc.). Subsequent attempts remain "pas bon" (no good). Numerous tweaks and hypotheses follow. Finally, two pages later, at around the 12th attempt, Yves notes the new oil viscosity ("Huile 47 V 20") and the new current ("$I = 3.6$ A"). And then: "Ça Marche!" (It Works!) At the bottom of the page, Stéphane writes, "en plus c'est vrai!! . . ." (also, it's true!! . . .).

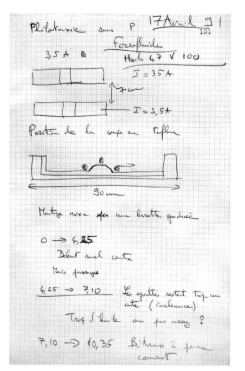 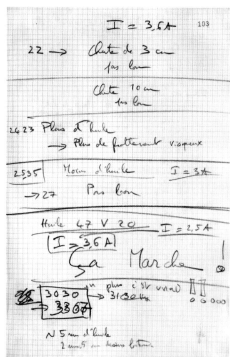

FIG. 13.5 Lab notes from Yves. On about the 12th try of the experiment, after changing the oil in the bath and the magnetic field, he wrote, "Ça Marche!" It works! Can you read his mood in the letter size?

The experiment was tricky: the plate had to be perfectly horizontal, the pipette perfectly centered, the oil viscosity just right, and the gradient of the magnetic field precise. While it was fairly easy to obtain patterns with simple spirals, it was difficult to produce several spirals at the same time. "In my memory, the experiment only worked three times," Stéphane recalled. "But the three times it worked, it was really impressive."

The experiment showed the spontaneous formation of Fibonacci spirals, with a divergence angle tending toward the golden angle. "It worked, but actually I did not understand why," Stéphane said. "The first spirals, (1, 1) and (1, 2), were easy to understand, as there are no other possibilities. But higher numbers remained a mystery. I must say, I was a bit skeptical."

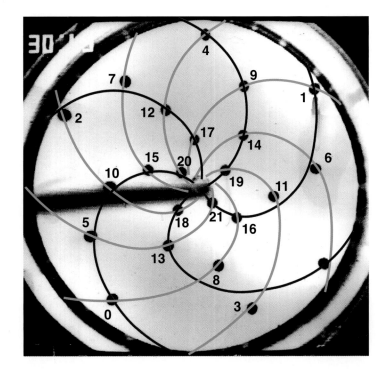

**FIG. 13.6** The lab camera featured new technology: a timer (there is a time stamp at top left). Note that the droplets are not quite perfectly placed. Would you erase the stuck droplet to make the image look nicer, or not?

There was no denying the sheer visual beauty of it, though. People loved watching the droplets appear and do their spiral dance. The experiment and its accompanying theoretical explanation captured the imagination of both fellow scientists and the public. The first short paper Stéphane and Yves wrote about it was well received, quickly accepted, and published in the most prestigious physics journal, *Physical Review Letters*. They also wrote an article for nonscientists and gave many popular talks.

A few years later, Ian Stewart's article about phyllotaxis in *Scientific American* ran an imaginative explanation of their work. (This is the same article quoted in chapter 12.) He wrote the piece as a dialogue between a shepherd, Grimes, and a gooseherd, Bumps, who are trying to understand why Fibonacci spirals occur:

Grimes snapped his fingers. "You're telling me that the numbers arise through some mathematical mechanism? Physics, or chemistry, or—"

"Dynamics," Bumps said firmly.

"Has somebody actually explained how plant growth might yield Fibonacci numbers?" Grimes asked.

"Well, lots of people have suggested many different kinds of answers. But for me, the most dramatic insight comes from Stéphane Douady and Yves Couder of the Laboratory of Statistical Physics in Paris. They recently showed that the dynamics of plant growth could account for the Fibonacci numbers—and much more."[4]

## Numerical Simulations

While Yves was perfecting the magnetic fluid experiment, Stéphane was exploring a complementary approach to explaining the physics of phyllotaxis: computer simulations. "It wasn't too difficult to imagine a numerical program reproducing this process," he recalled. 'But what were we looking for? Something that would select a pattern with two numbers—the Fibonacci parastichies in both directions. But I had only one parameter: growth."

The growth parameter was easy to see in the magnetic fluid experiment. "As the drops drift outward, the ratio of the distance traveled in one period versus the size of the central zone gives you a geometric ratio," Stéphane explained. "If you wait a long time between drops, you get a large growth parameter. The particles drift away before the next one appears, and you can see the drops are repelled only by the previous one and form a straight line. But when you reduce the time between drops, they start to spiral. When you reduce the time more, you get a system with five clear spirals and (5, 8) parastichy numbers."

That was nice, but growth alone wouldn't work. "I needed another parameter. Otherwise I was going to find a *family* of solutions—like (1, 12), (2, 11), (3, 10), (4, 9), (5, 8), (6, 7), all with 13 orthostichies—but not one *single* solution," Stéphane said. "So I stopped." The search for a second parameter would preoccupy him for a long time.

What happened next provides a nice example of how science works. Sometimes it just takes a nudge in the right direction. Stéphane recalled: "One day Yves asked me, 'What's happening with your numerical program?' I replied, 'Mm, won't work.' He said, 'Really? Just try it!' So I did. The first run gave the results (13, 21). It was perfect. I was shocked. Then I worked on the program a lot. I did all the simulations for all the cases, with diagrams. I recomputed the Van Iterson diagram." Once again, a model's power to match the Van Iterson diagram served as an important test.

His numerical results, shown as black triangles in figure 13.7, match Van Iterson's all-important geometric diagram—turning it sideways and replacing the r-axis (rise) with the dynamical parameter $G$. (For more on this parameter, see the appendix.) Not all of the branches connect, however. Only those

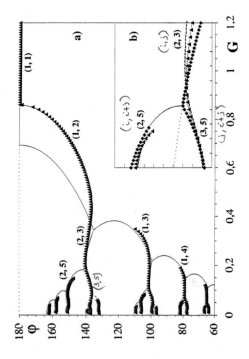

**FIG. 13.7** Stéphane generated this figure from computer simulations. Only the Fibonacci lattices remain connected—from (1, 1) to (3, 5), for example—starting at 180° (opposite leaves) at the top, and oscillating down to 137.5° (golden angle leaves) at the bottom. (For more details, see the appendix.)

branches corresponding to the Fibonacci lattices remain connected throughout. The diagram provides proof that, if plants always start from a large parameter $G$, and if $G$ is reduced slowly enough, only Fibonacci lattices will be observed. (Exceptions occur when starting from $G$ that is too low, or when $G$ decreases too quickly.)

## The Trick

But Stéphane still wasn't satisfied. Why did his numerical simulations work right when the parameter was reduced, yet Van Iterson's tree could still lead anywhere? Stéphane had hit upon a puzzle. "In the Van Iterson tree, you could start with any pair, and when you reduced the parameter, the diagonal would collide and two new solutions would appear," he said. "You could choose either one side or the other. From there you could generate all the pairs of numbers you wanted—not just specifically Fibonacci ones."

That's when Stéphane discovered the article by Levitov, talking precisely about the broken choice at the bifurcation. (Recall that bifurcation is a kind of physics fork in the road: a small change in parameter values causes a sudden shift in behavior, usually creating two options). Levitov had shown in his theoretical vortex models that bifurcations in the Van Iterson diagram were, with one exception, not symmetric. Where *two* new valleys appear in Van Iterson's picture, in Levitov's model there was only *one*. This creates what is called a "broken bifurcation," where only one side is chosen.[5]

Now when Stéphane was reducing the growth parameter in his simulations, he saw that the solution always shifted to the "good side." The two branches were not symmetric, and that was a revelation. "That shift is what pushes the Fibonacci numbers," he said. "Broken bifurcation was what made Fibonacci work!"

Now the Van Iterson tree had been "pruned" to show separated branches. But on these branches, signs of bifurcation still appeared, as shown in Stéphane's plotting of points on the Van Iterson diagram (fig. 13.7). As Ian Stewart so nicely described it: "The main branch runs close to a divergence angle of 137.5 degrees, and along it you find all possible pairs of consecutive Fibonacci numbers, in numerical sequence. The gaps between branches represent 'bifurcations' where the dynamics undergo significant changes."[6] Where there are bifurcations, you see a branch that the primordia cannot follow, just as Turing wanted to show.

**So Why Does It Work?**

At this point, Stéphane felt he had made some progress on the Fibonacci problem. And then, once again, a colleague pushed him to the next level. When Stéphane revealed his latest diagrams, the physicist Vincent Hakim said, "What you're showing me is nice, but why does it work?" He pushed Stéphane for an explanation.

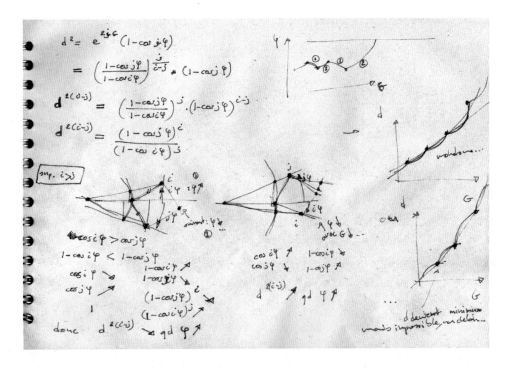

**FIG. 13.8** A page from Stéphane's notebook as he was playing with ideas of asymmetric bifurcation, sketching two diagrams of the meristem center. A piece of its circular border forms a "pie slice," while the beginnings of parastichies in two directions form a square. Why would the square be tilted?

"So I went back to take another look at the symmetry argument," Stéphane recalled. If you draw Van Iterson's tree, he said, you can see that the phyllotaxis mode before the bifurcation is not symmetric, except for mode (1, 1). When one parastichy number is larger than the other, the diagonal is always shifted to one side. Where you reach the bifurcation, the older primordia push the new primordia "to the correct side," breaking the possibility of their going to the other side. But this was still the same symmetry argument that Schwendener and Levitov had already made.

To understand Stéphane's new dynamical insight, you have to look at when and where the forbidden branch starts to reappear on Van Iterson's tree. In the Levitov model, the new solution always emerges after the threshold of crossing a "mountain pass" arc. The mountain pass corresponds to the highest repulsion, when the lattice is in a square state (with maximum space between primordia). But then Stéphane found that the American mathematician Irving Adler had proposed a different threshold: where the new parastichy is vertical.[7] Stéphane wondered where exactly this threshold appeared in his own simulations.

When he looked again, Stéphane saw that the threshold of his "pruned" branches appeared *between* the two limits proposed by Adler and Levitov. "That was when I found a dynamical argument," Stéphane said. It struck him that in nature, not all of a plant's organs are created at the same time. "They're created *one by one*, just as Hofmeister told us. And that makes a big difference," he said. "You have to put your new element where there's a minimum, in a hole. And at a bifurcation, you have a branch that is forbidden, because the position that the regular geometric lattice suggests is not a minimum. The new primordium wouldn't stay there. It would roll down to a lower place!"

In doing his computations, Stéphane found a general formula for determining a place of dynamical stability. "That was quite a surprise for me," he said. "I guessed there would be a general formula, I tried to write it down, and right away it worked. I remember being quite amazed at my table, looking at this beautiful result."[8] He wrote an article and sent it to a journal on dynamics and chaos. But disappointment followed. The reviewer rejected it, saying, "I don't see any dynamics here." Later the paper found its place, published in the book *Symmetry in Plants* by Roger V. Jean and Denis Barabé.

In figure 13.10, Stéphane shows the three different possible thresholds for where a new primordium emerges. The dotted lines indicate Adler's threshold, where the new parastichy would be vertical.[9] Levitov's threshold, where the lattice is square (favored by Van Iterson himself), is indicated by the blue dots. Stéphane's threshold, indicated by the short transverse lines, is where the arc of solution becomes dynamically stable. Usually, Stéphane's threshold is between Adler's and Levitov's, but not always for very asymmetric patterns. This shows that the dynamical threshold is actually independent from the two geometric thresholds—Adler's vertical one and Levitov's square one.

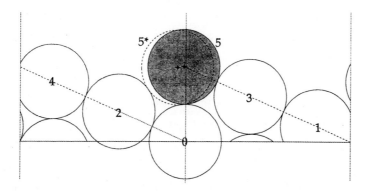

**FIG. 13.9** To fulfill the requirements of dynamical stability, disk 5 needs to find a place that is not only a minimum but the *best possible* minimum. (For more details, see the appendix.) Do you think that disk 5 is at a minimal position?

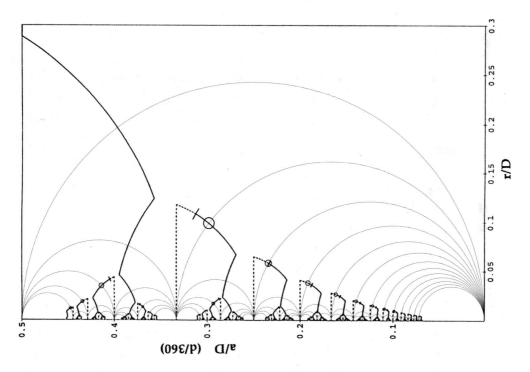

**FIG. 13.10** Another version of Van Iterson's tree, this time pruned to show Stéphane's threshold of dynamical stability. This threshold—marked by the short transverse lines—is distinct from Adler's threshold (shown by the vertical dashed lines) and from Levitov's (at the crossings with the half circles). Are Stéphane's located in any special place?

### The Missing Parameter

In the end, the lesson Stéphane learned was this: the Hofmeister model was truly different. Hofmeister's one-by-one approach to plant growth led to a model quite distinct from those of Adler and Levitov, who assumed a regular lattice had already been built. The new model advanced by Stéphane and Yves took a fresh approach, putting each new primordium in its optimal place—*locally*—and from that the global pattern emerged.

"What really makes this model work for me is that you construct it dynamically, step by step," Stéphane explained. "You can imagine lots of nice geometric solutions, but you still have to build them one element at a time, and that process imposes a particular solution. That made me realize, yeah, there's continuity of growth. You start from the seed. Then to place the leaves, you have to decrease the size of the primordia with respect to the central zone. You decrease again to form the bracts and the outside rim of the flower. Then you increase again to finish the flower center. It is this whole history that gives you Fibonacci numbers in a sunflower."

And that vision led to a final insight. "That's when I realized that the missing parameter was *history*," Stéphane said. "In morphogenesis, you're not building the best solution. You're building the best you can, step by step. When you see an element of a plant, what you see is the result of a whole development and growth history. All the steps that followed in the growth are visible, and they impose their markings everywhere."

This insight led to another paper coauthored with Yves, this time bringing in what is known as the Snow and Snow rule: a new primordium appears not just where it has enough space, but also *when*. The rule's creators were Oxford botanists Mary Pilkington Snow and George "Robin" Snow, who collaborated starting in 1931 on several notable phyllotaxis papers, based on their painstaking experiments on plants to better understand why and where leaf primordia originate.

And so, over the course of a single year, 1996, Stéphane and Yves published three major papers on phyllotaxis. The first was based on the Hofmeister model, describing their experiment with magnetic droplets. The second was based on

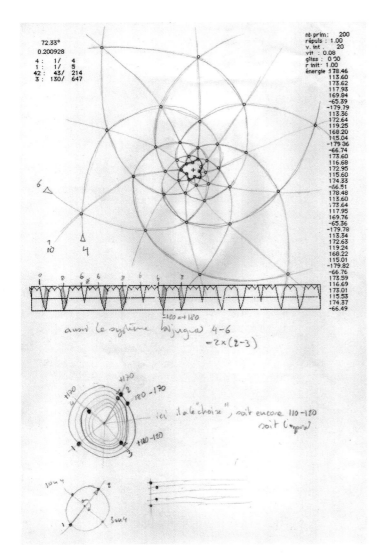

**FIG. 13.11** This is Stéphane's first computer simulation applying the Snow and Snow rule: a new primordium appears not just where it has enough space, but also when. (For more details, see the appendix.) Can you count the number of spirals?

the Snow and Snow model, in which a primordium can appear not only where but also when there is enough space. The third paper merged their prior work, explaining that the continuous transitions of spirals in a plant—its history—are what is needed to explain the occurrence of Fibonacci numbers in plants. "After that," Stéphane said, "we were exhausted."

At that point, he took a break from phyllotaxis. "I didn't think we could do much more," he said. "Anything else seemed to add complications that weren't necessary or compelling." There was one problem that still interested him, however. One day after Stéphane finished giving a talk, an audience member asked, "What do you do with the case of corn, where the number of vertical spirals is always an even number?" Stéphane replied, "No idea yet!" Later, he realized that corn kernels have to be grouped in pairs, with one primordium giving rise to two horizontal seeds. Beyond that basic pattern, the numbers of spirals looked strange—either identical, as in a whorled case, or differing by only one. But corn is the product of thousands of years of selection by indigenous peoples of Central America. Unable to survive in the wild, it seemed to Stéphane a "monstrous" case.

Progress on corn and other phyllotaxis oddballs would have to wait until Stéphane met Chris.

CHAPTER 14

# Zigzag Fronts in an Artichoke

> Flash!
> All
> at once
> Chris saw the
> power in zigzags—
> it was a eureka moment.
> An artichoke can take you unexpected places.

Following his dramatic experiment with magnetic droplets, Stéphane felt he'd reached a Fibonacci plateau. Moving beyond it would have to wait until he met Chris, the mathematician who would become both his collaborator and his friend. Chris's breakthrough—he calls them *zigzag fronts*—created a new way to understand irregularity in plant spirals, while finding Fibonacci order within that complexity. It even turns out to be a radically simple explanation for the origin of Fibonacci in plants.

The thing is, real-life plants pose complications that scientists mostly either ignored or explained away. "When you start counting parastichies on a sunflower, there's a problem," Chris said. "Even on the ones that look the most regular, you'll try to trace a parastichy and get stuck. You have to choose between two paths." These forks in the road appear at transitions, places where the number of plant spirals increases or decreases as the plant grows.

**FIG. 14.1** At Smith College, Chris helped create a popular exhibition on phyllotaxis. An expanded version opened in Geneva, Switzerland, as Jardin de Maths (The Math Garden). Do you see any spirals?

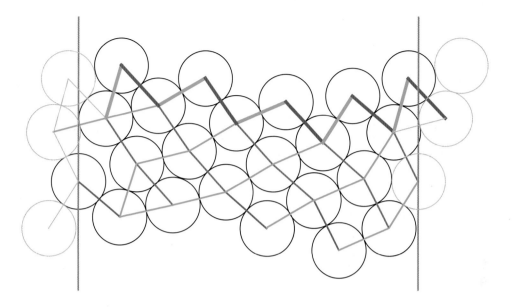

**FIG. 14.2** In Chris's simulations, the parastichies were rarely straight. At the upper edge (near the meristem), they formed a broken line he calls zigzag fronts (shown as the bold green and red segments). Do you see different patterns, *locally*?

There at these confusing junctions, Chris saw zigzagging chains surrounding small rhombuses, but also triangles and pentagons. In a flash, he realized that the triangles and pentagons actually encode the increases and decreases in spirals that occur at transitions. "To me, these fronts explain a lot," Chris said. "They're made of little pieces of parastichies, little local pieces that don't need the whole structure to live. Because they are little and local, they are more nimble!"

Why did Chris suddenly see these subtle geometries where others had not? He credits his background in mathematical topology for paving the way—and maybe even his familiarity with musical chord progressions, given that he's a serious jazz pianist. His discovery of zigzag fronts opened doors for other scientists as well. He gave Stéphane the tool he needed to tackle corn, which for many years had been his phyllotaxis "monster." As Stéphane developed his concept of quasi-symmetry, irregular plants like strawberries and acorn caps also yielded up their secrets, providing a new vision of phyllotaxis itself.

### The Back Story

In his academic career, Chris landed on phyllotaxis completely by accident. Growing up, he attended high school in Algeria, where his father worked as an engineer. Chris then spent a year studying for the French *grandes écoles*, the premier engineering schools in Paris. The math curriculum included topology, a subject that had a profound influence on his future. "At the time, I couldn't even do calculus properly, but I could do topology," Chris said, recalling why engineering lost its charm. "I understood it immediately. Topology is a total abstraction of space, and I just loved that. It's the core of my mathematical being."

From his undergraduate studies in Paris, Chris made a leap to Santa Cruz, getting his master's in mathematics at the University of California. A decade later, he returned there as a visiting professor. While Chris was teaching a graduate seminar in dynamical systems, his student Scott Hotton proposed something unorthodox: a project relating the fractal Mandelbrot set to plant spirals. "I was both fascinated and a little suspicious," Chris said. "There's a lot of stuff around fractals that feels pseudoscientific because it's flashy and people are drawn to the beautiful pictures." But Hotton's project turned into a PhD dissertation, and from that point on, Chris was hooked on phyllotaxis.

Taking a professorship at Smith College, Chris began researching the mathematics behind plant patterns. Over many years, his work would link him to this book's coauthors—Stéphane the physicist and Jacques the biologist—and that is how *Do Plants Know Math?* came into being. Pursuing his research, Chris began looking at how each primordium forms with its two nearest neighbors. Together with Hotton and Smith colleague Pau Atela, he wrote a proof showing that a simple mathematical model could generate Fibonacci phyllotaxis.[1]

And what exactly was their proof based on? On a model created in Paris, by none other than Yves Couder and Stéphane. The problem was that this model was *too* simple: it assumed that primordia formed one by one at prescribed time intervals. The resulting pattern worked in some cases, but it ignored a large swath of plants with whorled phyllotaxis—those with two or more leaves at each level of the stem. Hoping to fix the problem, Chris wondered whether disk stacking might have some potential.[2]

First, he wrote a disk stacking code. He ran thousands of simulations, trying to get a sense of what happened as plants grew. "Something interesting came up," he said. In the study of phyllotaxis, scientists had always been obsessed with a fixed divergence angle. But in his models, he said, "I kept *not* seeing it. What I saw instead was *periodic sequences* of divergence angles. So I was really intrigued."

Looking at patterns that emerged in his simulations, he noticed that the periodic divergences came from squiggly lattices he called "rhombic tilings." These tilings break the dominance of the fixed divergence angles.[3] "This means you cannot really understand the pattern through divergence angles alone," Chris said. "You have to get a grasp on the pattern another way."

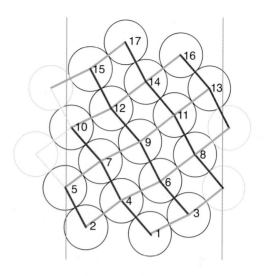

**FIG. 14.3** An example of rhombic tiling in Chris's disk stacking simulations. These show neither triangles nor pentagons, but instead crooked parastichies.

At this point in our story, we connect the dots to Jacques, who was doing research on plant biology at Harvard. In 2005, he invited Chris to work on a group project using computer-generated models to analyze plant pattern data. This was Chris's first real foray into applied math. There were no theorems to prove but instead a lot of code writing and concept developing. As its top priority, the group wanted to understand why the number of segments going up and down matched the number of parastichies *locally*. Starting from this question, Chris sought to explain Jacques's beautiful microscopic pictures.

Photos like the artichoke in figure 14.4, which clearly shows every primordium, are rare and difficult to produce. At this stage, the meristem measured only five millimeters wide! To prepare the specimen, Jacques had to painstakingly remove the surrounding bracts and make an imprint using dental putty. Then

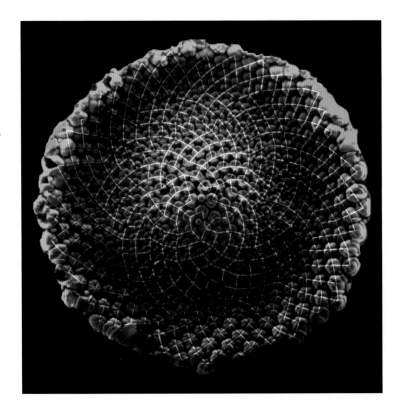

**FIG. 14.4** Jacques's scanning electron microscope photo of an artichoke meristem, with lines showing the parastichies. Each bump is a flower bud ready to grow out as a "hair" on the artichoke heart. Where do you see the number of spirals changing?

he sprayed the impression with metal particles to make it visible to the scanning electron microscope. (Just imagine doing all this without destroying the fragile florets.) Once Jacques had taken the micrograph using an SEM, Chris and his colleagues converted it into a computer model, shown in figure 14.5.

When Chris saw the artichoke through the eyes of a computer diagram, he had his eureka moment. In a flash, he realized that zigzag fronts contain information about not only parastichies but also their transitions, the places where spiral numbers increase and decrease as the plant grows. Thinking back to his discovery, Chris said that the idea of zigzag fronts seemed like low-hanging fruit, just waiting to be picked.[4] "But the moment when things clicked—when I knew, ah, this is it!—was when I realized how these fronts encode transitions." For a detailed explanation, we will take a closer look at the artichoke.

**FIG. 14.5** This computer-generated model of the artichoke meristem led to Chris's eureka moment—when he saw zigzag front transitions appear before his eyes.

Three zigzag fronts are shown in figure 14.5, depicted as three gear-like circles made up of red and green line segments in boldface. The outer, larger front developed first as the artichoke grew. It has 55 red and 34 green line segments. The middle front reveals that now the meristem has filled up, with the numbers of red segments decreasing to 21, while the 34 green ones remain. In the smallest front, closest to the artichoke's center, 21 red segments remain, while the green segments have decreased to 13. (Note that each part of a double segment is counted separately.)

These numbers of zigs and zags in the fronts—note that all are Fibonacci numbers—correspond to the plant's traditional parastichy numbers, as we will see in the simpler *Anthurium* pattern. But that is not all the information that zigzag fronts encode. Keep reading!

### Fronts, in a Nutshell

Change is always difficult. Making the shift to a new paradigm—a whole new way of seeing—is often a long and painful process, perhaps a bit like giving birth. In order to clearly explain the new zigzag front paradigm, we are going to sweep away all discussion of the stumbling that occurred along the way, along with the historical debris. Dear reader, we now ask you to forget nearly everything you have learned in this book up to this point, including divergence angles, Fibonacci numbers, the golden ratio, and even the Van Iterson diagram. Instead, our starting place will be the first observations of Wilhelm Hofmeister, the self-taught botanist and microscope wizard we met back in chapter 8.

It turns out that a good place to see zigzag fronts is the meristem, the microscopic tip of a growing plant observed by Hofmeister long ago. As we have mentioned before, the tiny new bumps that appear in the meristem are the primordia, tightly packed between the meristem's central region and the already existing primordia. The new primordium—call it a "child"—will fill the largest available space, fitting snugly between two previous primordia, its "parents." This new arrival has one parent on its left and one on its right. The sequence can be clearly seen in the SEM micrographs of a spruce (*Picea*) branch in figure 14.6, in which every primordium bump would turn into a needle.

 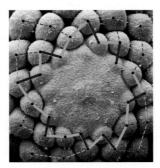 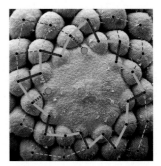

**FIG. 14.6** The meristem of a spruce. In the middle image, a front—shown in bold red and green lines—joins the parent primordia to the new children. At right, orange dots show where the two newest primordia choose the largest space available between two parents.

If we follow the lines that join parents to children around the meristem, we see the zigzag front emerge. All the "up-steps" (from parent to child) appear as bold green segments. All the "down-steps" (from child to parent) appear as bold red segments. These steps can be extended into spirals, the parastichies we know so well, running in opposite directions. As we will see in a moment, counting steps reveals a clever but somewhat subtle trick for counting parastichies. To see this more clearly, we will look at fronts drawn on an *Anthurium* flower spike that has been unrolled to show its patterns on all sides.

## Counting Parastichies Using Fronts

Looking at the photos in figure 14.7, we see how for the front at point *P* each red step connects one blue parastichy to the next. And here's the punch line:

The number of *red* steps, 5, matches the number of *blue* parastichies. And vice versa!

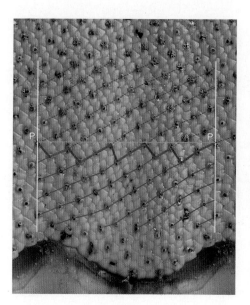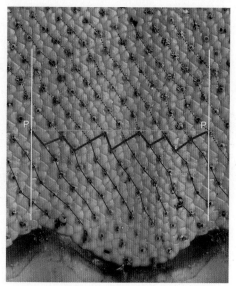

**FIG. 14.7** In this "unrolled" *Anthurium*, the zigzag front at *P* has 5 red steps and 8 blue ones, indicating (5, 8) phyllotaxis. The number of red steps matches the number of blue parastichies: 5. And the number of blue steps matches the number of red parastichies: 8.

To see how this works, we can understand the red steps as separating the blue parastichies from one another. In the same way, the front's 8 blue segments separate the red parastichies by stepping from one to the next. (Note that each part of a double blue segment is counted separately.) It turns out that this method of counting parastichies is more versatile than our old approach: fronts and their parastichy numbers are still there to be counted even when parastichies are hard to trace. This is particularly useful where parastichy numbers change.

## Fronts and Transitions

Recall that transitions are places where the number of plant spirals increases or decreases as the plant grows. If you're counting spirals on a strawberry or a daisy, transitions are the places where you meet a dead end. Zigzag fronts are particularly powerful in understanding transitions, as shown in figures 14.8 and 14.9, using the flower spike of a peace lily.

**FIG. 14.8** Here, the flower spike of a peace lily shows a *triangle transition*, where an additional parastichy appears. The place where this happens, marked by a triangle, is where the blue parastichy that rises from the bottom left splits into two.

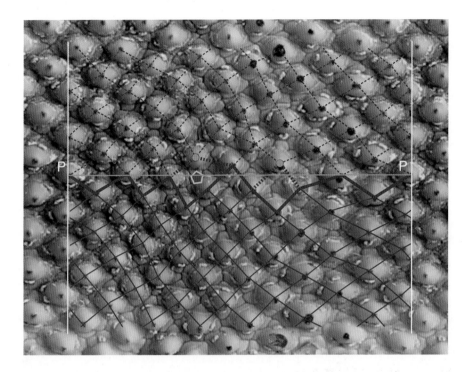

**FIG. 14.9** This image of another peace lily shows a *pentagon transition*, where the number of parastichies decreases by one. The place where this happens, marked with a pentagon, is where the blue parastichy coming from the bottom right dies out.

At this juncture, let's zoom in closer to see how these triangles and pentagons form—while not neglecting the rhombuses, which have their own role to play. When a new primordium appears, it sits on its two parents, one to the left and one to the right. Typically, the two parents themselves share a parent, the three forming a *V*. Together with the new child, these three primordia form a rhombus, which simply swaps a pair of down-up steps for a pair of up-down ones. Therefore it does not alter the zigzag front's number of ups and downs, and the local parastichy numbers remain unchanged, as shown in the center photo in figure 14.10.

In some cases, however, a rhombus can't take shape. This happens if the zigzag front becomes too flat, such as when the primordia are shrinking. Then the *V* formed by parents and their common grandparent opens up too much, and

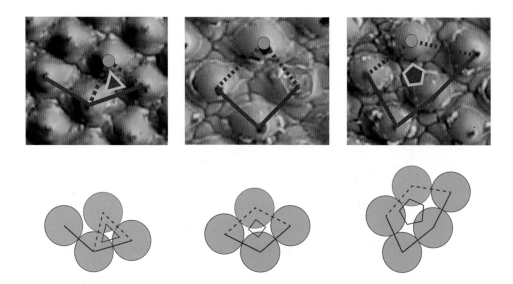

**FIG. 14.10** These photos and their accompanying sketches show three types of transitions in more detail. The triangle at left adds a parastichy, the rhombus in the center does not change the zigzag front or the parastichy numbers, and the pentagon at right subtracts a parastichy.

the new primordia can't keep the same parents without overlapping the grandparent in the middle. Instead, the new primordium sits on the side of the *V* that is the most horizontal—following Hofmeister, as this is where there is the most space. This creates a triangle and an additional parastichy, as seen in the peace lily in figure 14.8. (For a close-up view, see the left-hand photo in figure 14.10.) If you try to trace the rightward-sloping parastichies, you can see where the split occurs at the triangle.

The converse also happens, where primordia in a zigzag front get crunched together. The new primordium cannot find a place between its two logical parents, as it is now pushed out of place by a primordium next to one parent. It has no choice but to sit between this side primordium and the other parent. These primordia form a pentagon. This time, the new front has one less step, and the parastichy number decreases by one. In figure 14.9, you can trace where a leftward parastichy comes to a dead end at the pentagon. (For a close-up view, see the right-hand photo in figure 14.10.)

## Fibonacci Transitions (Finally) Explained

The miracle is that, armed with only these basic concepts of zigzag fronts, we can now understand the emergence of consecutive Fibonacci numbers of parastichies.

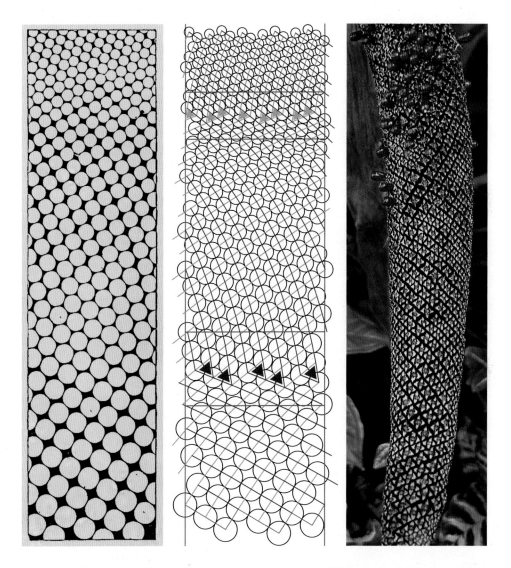

**FIG. 14.11** At left, a drawing of consecutive Fibonacci transitions from 1900, starting with (3, 5) at the bottom, then changing to (5, 8) and (8, 13). At center, a modern simulation of the same. At right, a similar smooth transition in the spadix of *Anthurium digitatum*. Where do the parastichies become flatter?

As the tip of the stem grows, each step of the zigzag front must stretch to reach around the stem, a situation that tends to create triangle transitions. If the pattern is regular enough, these transitions will occur in an orderly way. As an example (see fig. 14.12), let's start with a front having 5 down-steps and 3 up-steps, and then see what happens as the plant grows. Because the front begins and ends at the same point on the stem, the 3 down-steps will be, on average, steeper than the 5 up-steps. As we've observed, triangles must lie on the flatter steps. So in this case, as the plant's growing tip adds primordia, there will now be 5 triangles on the down-steps, adding a total of 5 *new* up-steps. This makes the new total of up-steps

$$3 + 5 = 8$$

In the meantime, the 5 down-steps have just been replaced by the same number, 5, of (steeper) new down-steps. In other words, from a (3, 5) front we have transitioned to an (8, 5) front—revealing the law of Fibonacci phyllotaxis.

So now we've seen how, starting with one pair of Fibonacci parastichy numbers, we can move on to the next pair. But what about the original starting point, the parastichy numbers (1, 1)? It turns out that many plants emerge from seeds with just one leaf (or cotyledon) at the base of the stem. In these monocot

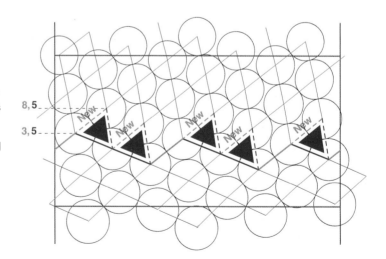

**FIG. 14.12** Here, a (3, 5) zigzag front becomes (8, 5) after 5 triangle transitions. Each triangle rests on a red segment and creates a new green segment. The triangles can be seen splitting the red parastichies into two as they rise up from below.

plants, the next leaf emerges a little higher on the stem and on the opposite side, initiating the (1, 1) pattern. Other plants—dicots like sunflowers—start with two opposing cotyledons and a (2, 2) pattern. Very often, this breaks into a (2, 3) pattern and continues along the Fibonacci sequence.[5]

In figure 14.13, we see the whole zigzag front framework in a real plant, proving that the concept is no mere mathematician's daydream. When Stéphane cut the leaves off an ornamental cabbage, their traces beautifully revealed Fibonacci transitions from (1, 1) to (8, 5). Fronts of successive Fibonacci parastichy

**FIG. 14.13** It's rare for a photo to capture a full set of Fibonacci transitions. Here, these transitions appear in an ornamental cabbage with its leaves cut off. How many triangles are there at the successive transitions?

numbers and triangles neatly showed the transitions, just as they appear in computer simulations.

## Coda: Discovering Quasi-symmetric Phyllotaxis

Why are zigzag fronts helpful? For one thing, as we have seen, they offer an easy, visual explanation of why Fibonacci phyllotaxis is so prevalent. These fronts also show the growth conditions necessary for Fibonacci transitions to occur. Violate these conditions and you get patterns that scientists long ignored because they didn't fit their models. This is how zigzag fronts led Stéphane to discover quasi-symmetric phyllotaxis.

### STÉPHANE DISCUSSES CORN AND OTHER MONSTERS

When Chris told me about his zigzag fronts, I was immediately enthusiastic. Because there was one problem still in the back of my mind: how to deal with irregularities.

In the numerical sunflower simulations we did in our lab in Paris, we saw many permutations of the divergence angle. Irregularities and quick transitions had also appeared in my first numerical simulations. But I'd overlooked them because they didn't fit the Van Iterson tree, which was based on regular patterns. I'd thought about investigating these irregular cases, only I didn't know how to approach them. Chris's fronts were the tool I was waiting for.

Another surprise came almost a decade later, when Chris told me that if disk size decreased too quickly, the usual Fibonacci patterns did not work. "How do they look?" I asked. "A bit chaotic," he said. "How chaotic?" I asked. "Show me."

When I looked at examples from his simulations, I found that they were exactly like the modes I had first observed in corn. In corn, one primordium gives rise to two horizontal seeds. I found modes of nearly equal numbers of parastichies, meaning equal to or differing by one.[6] Equal numbers could fit whorled phyllotaxis, those plants with more than one leaf at each level. But "not-quite-whorled" had not yet been described, and I'd never published my corn observations.

After that, I began looking for all the strange phyllotaxis cases, in particular those with nearly equal numbers, which I then called quasi-symmetric. I found some odd Australian plants with flower spikes, *Banksia*.[7] Then I found *petai*, the "stinky beans" that grow wild in the Indonesian tropical forest. With my list consisting of one phyllotaxis monster, one Australian genus, and one wild tropical plant, my quest wasn't looking very successful.

But the fact that Chris was finding the same patterns was exciting, so I started looking more closely.[8] I remembered that when I first started to work on phyllotaxis, I'd looked at the case of strawberries but couldn't figure it out. And so I'd forgotten about it.

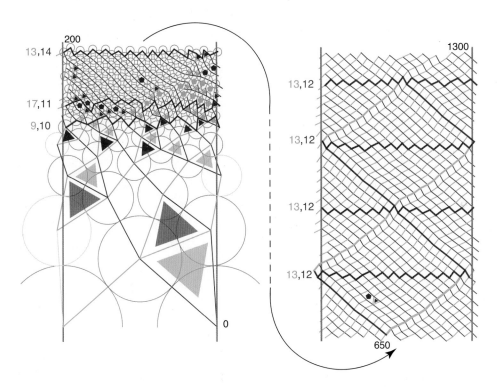

**FIG. 14.14** Quasi-symmetric patterns occur when disk size decreases very fast. After 36 disks, the size was kept constant, resulting in a cloud of triangle transitions. Do you see the mix of red and green triangles occurring at the same levels? This led to the unusual tiling (13, 14) at top left, which later stabilized—after 700 disks—to (13, 12) at right.

But now, taking another look revealed that strawberries presented exactly the same quasi-symmetric case. There were also plants in the Araceae family that fit, such as the white peace lily, with its unusual spadix. The most recent case I've found is acorn caps. So it turned out that these modes are more common than I had originally thought, and that we could indeed define two main families of phyllotaxis modes: the Fibonacci type—including those starting with numbers other than (1, 1)—and the quasi-symmetric type, with spiral numbers close to each other.

As a new way to see plant geometries, Chris's zigzag fronts helped explain the prevalence of Fibonacci numbers in phyllotaxis in very simple terms, in addition to shedding light on some of the most puzzling phyllotaxis misfits. In the next chapter, we'll move on to fractals, the mathematical patterns that initially made Chris somewhat queasy. He and Stéphane will debate an interesting question: Are phyllotaxis patterns truly self-repeating, or are they not?

CHAPTER 15

# Self-Repeating Patterns in Plants (Perhaps?)

Don't
you
wonder,
Broccoli,
whether you repeat
yourself? I mean, really repeat
your *self*? How much self-similarity is enough?

Somehow, we humans find fractals endlessly appealing. It's worth recalling that fractals—self-similar patterns of infinite complexity—were not named or described as a common family until the mid-1970s. Before fractals, people lacked tools for understanding complex irregular shapes. Some complex curves were known, such as the brachistochronic curve of fastest descent and the logarithmic spiral seen on a pinecone. But except for a few outliers, these remain globally smooth. By contrast, fractal pictures tend to look like ferns: very pointed and convoluted. They are called "self-similar" because, for example, a part of the leaf looks like the whole leaf (see fig. 15.7). Yet while plants display many fractal (or fractal-like) structures, they did not inspire the discovery of fractals. So where exactly do plants fit in?

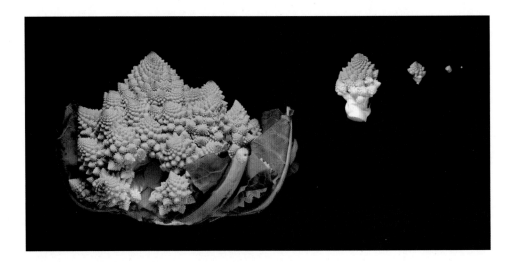

**FIG. 15.1** Romanesco broccoli: fractal or not fractal?

While this chapter was being written, a debate arose between Stéphane the physicist and Chris the mathematician over certain questions: Are plants truly fractal? Is it enough for them to be self-similar? And what about the self-similar Van Iterson diagram we've seen so often in this book—does it count as fractal too?

Before we explore the debate further, let's go back into history. One of the first forays into complex objects was accidentally made by Georg Cantor (1845–1918), the German mathematician who created set theory. Cantor tackled the tricky topic of infinity in sets of numbers: he rigorously defined, distinguished, and manipulated them, asserting that there is not just one type of infinity but a whole hierarchy of infinities. His results were often counterintuitive, and at times they conflicted with his devout Christianity. More than once, rival mathematicians viciously attacked his ideas, and Cantor (who possibly had bipolar disorder) ended up in a sanatorium, suffering from depression. In response to his critics, he wrote:

> The fear of infinity is a form of myopia that destroys the possibility of seeing the actual infinite.[1]

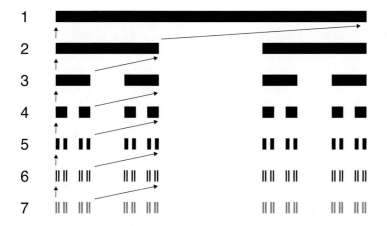

**FIG. 15.2** This graphic shows the first seven steps for constructing the Cantor set, named after the German mathematician Georg Cantor.

In 1883, Cantor demonstrated the unintuitive nature of the size (cardinality) of a set, using an example that was later named after him: the Cantor set.[2]

In figure 15.2, Cantor starts with a single line segment. For the next iteration, he cuts the segment into three equal parts and then removes the central one. Using the resulting two segments, the same process is repeated, giving rise to four segments, and so on. Ultimately, the Cantor set leads to an infinite number of disconnected points, called Cantor dust.

After many iterations, the segments become so thin that they appear to vanish. Cantor could show, however, that even though no segments remained at the end, there were as infinitely many points in the final set as there were in the original line segment. This is a typical counterintuitive result for infinities of sets of numbers—in this case the infinite number of points on a line segment.[3]

Cantor's work was also a precursor to dynamical, iterated transformations. These show how a simple process, repeated again and again, can create a complex object. The same idea lies behind the familiar fractal object known as a Sierpiński carpet: cut a square into a three-by-three grid, remove the central square, repeat the process for each of the eight remaining squares, and so on. (The carpet was discovered in 1916 by Wacław Sierpiński [1882–1969], a prolific Polish mathematician.) Here the connectedness is preserved—the object can still be used as a carpet. Even if there are holes everywhere and the surface vanishes before our eyes, the carpet remains in one piece. And in fact, the concept has been applied to engineering as a way to reduce weight while preserving rigidity.

Another similar example is the Koch snowflake:

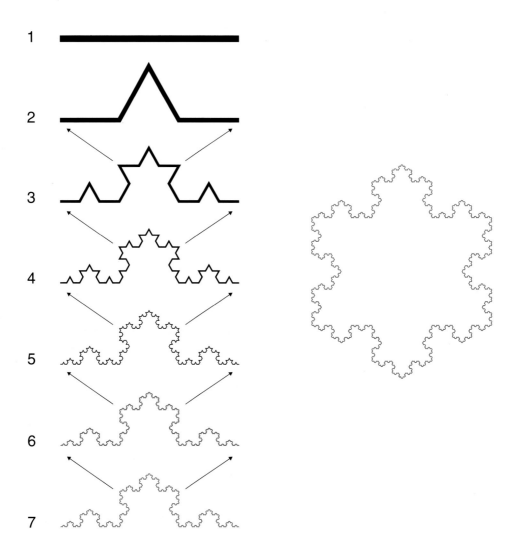

**FIG. 15.3** Here we see the construction of a Koch curve, an early 1904 fractal named for mathematician Helge von Koch. Join three of the resulting curves together, and you get a Koch snowflake.

Like Cantor, Koch begins with a line segment, then cuts it into three parts and removes the central part. But Koch adds the top of an equilateral triangle—whose side length matches the length of the section removed—to keep the curve connected. Again, this process is repeated for each new segment. As seen in figure 15.3, after the sixth iteration it becomes hard to distinguish the changes. The limiting result is a self-similar picture, with each fourth of the set being a copy of the whole set, rescaled by 1/3. The seventh iteration starts to look more like a snowflake, or perhaps the profile of a cauliflower.[4]

## Enter Mandelbrot

Not until Benoit B. Mandelbrot (1924–2010) came onto the scene, however, did fractals really capture the public imagination.[5] Born in Warsaw, Mandelbrot left Poland with his family in 1936 to join his mathematician uncle living in Paris. Given that the family was Jewish, the move may well have saved his life.

After getting his PhD in France, Mandelbrot joined a group of iconoclastic researchers at IBM in the United States, working on the "noise" that interfered with computer data being sent over telephone lines. When Mandelbrot graphed the noise data, he saw self-similarities that reminded him of work done by Gaston Julia, who had been Mandelbrot's math professor in Paris. Julia's nose had been shot off in battle during World War I, and afterward he wore a black leather strap across his face. He forged ahead undaunted, however, publishing his landmark paper in 1918 on iterations of functions of complex numbers. At that time, there was no way for Julia to illustrate the behavior of his iterated functions. But in 1961, Mandelbrot could do it, using state-of-the-art IBM computers developed in the wake of Turing's revolution. Mandelbrot ran millions of iterations.

In 1967, Mandelbrot published his paper "How Long Is the Coast of Britain?"[6] His surprising answer was, it depends. If you measure the self-similar wrinkles along the coastline down to the microscopic level, then the coast can approach being infinitely long (not unlike the Koch snowflake). In 1973, he wrote an algorithm that generated pictures of natural landforms, marking the first computer-generated "fractal pictures."

But not until Mandelbrot gave the concept a name did the field really take off—for as he put it, "to have a name is to be."[7] He coined the term "fractal geometry" in 1975, based on the Latin for "fractured." That same year, his book *The Fractal Geometry of Nature* was published in France, and soon the whole world was talking about fractals. Mandelbrot's central idea is that nature is full of fractal forms, not just in coastlines but in mountains, rivers, hurricanes, lightning bolts, and trees.

Mandelbrot created his fractal pictures by repeating the same process over and over again, just as in a dynamical system. These infinitely detailed pictures inspired not only artists but also mathematicians. One iconic image is of the Mandelbrot set—named after Mandelbrot by mathematician Adrien Douady (Stéphane's uncle). In turn, Adrien Douady had a fractal image named after him, the Douady rabbit.

Well then, what about our phyllotaxis friends like sunflowers? Mandelbrot argued that self-similarity alone was not enough to characterize fractals. He defined most fractals as having a noninteger (fractal) dimension, a measure of complexity represented by an exponent between 0 and 2 in the flat plane.[8] This exponent describes how the object grows in size at larger and larger scales. For a line, its length $l$ grows exactly at the scale $s$ we are observing, so $l = s^1$, and the exponent is 1. For a fractal curve, as its length increases it becomes more and more convoluted and locally dense while remaining in the flat plane, so its exponent is larger than 1. If it completely fills the plane, then its size grows like the surface, so $l = s^2$, and it has an exponent of 2. The Cantor set, a disconnected "dust," has a dimension of about 0.631, for example, while the Koch set has a dimension of about 1.26. And the spirals on sunflowers? Although the logarithmic spiral model of their parastichies is self-similar, it has a dimension of 1 because the spiral makes a smooth line. So perhaps sunflowers are not a great candidate for a fractal.

Some of the abstract self-similar structures generated by mathematicians come out looking like ferns. And in fact, many plants exhibit a natural fractal structure. So why didn't Hofmeister's drawings of plant architecture inspire discoveries in iterations? Why was the essentially fractal structure of plants not recognized sooner? Perhaps people tend to overlook the familiar forms around

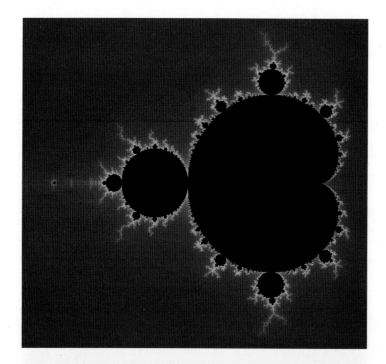

**FIG. 15.4** This iconic fractal image is known as the Mandelbrot set. The border of the Mandelbrot set is so dense in twists and turns that it has a dimension of 2.

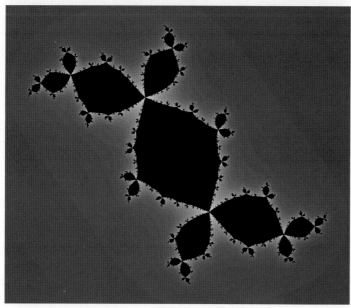

**FIG. 15.5** The Douady rabbit is named after the French mathematician Adrien Douady, Stéphane's uncle. Note its rabbit ears. A Julia set, it has a dimension of about 1.39. (For more details, see the appendix.)

them. Moreover, once fractals became popular, some mathematicians (including Chris) insisted that plants cannot be perfect fractals because their iterations of self-similarity are not infinite.[9] Take a look for yourself and draw your own conclusion. The photos in this chapter provide examples of near-perfect fractals in the leaves of a bur chervil plant and in Romanesco broccoli.

**Stéphane defends the fractal view:** I think it's unfair to say that plants are not perfect fractals because they fail to display infinite results. Most important is the *process*, the iteration of a simple transformation. As Turing brilliantly demonstrated, this is what allows us to understand the internal logic of shape. In a similar way, Turing believed that repeating the simple mental steps of computation would allow us to understand the nature of intelligence. Recent developments in artificial intelligence indicate that he was right.

Another reason for defending plant fractals is this: fractal enthusiasts like to measure the "fractal dimensions" of anything, often using free software—even if the object they end up with is not truly self-similar or fractal at all, just fuzzy. In the case of plants, it's definitely worth doing the measurements.

In figure 15.6, Stéphane's photos show the bur chervil's fractal (or fractal-like) structure in more detail. At bottom right, the successive order of the leaf shafts is numbered from largest to smallest.[10] The largest shaft, 1, is colored red; the smallest, 7, is colored blue. In the series of photos at the top, we see how the corresponding leaf parts are self-similar: 2 is similar to the top of 1, and so on.

## A Botanist Who Changed Math

Up to this point, we have seen many examples of scientists who used math to make new discoveries about plants. But one botanist made the reverse journey, using plants to create a new approach to math—based on repetitive forms that behave like fractals. Aristid Lindenmayer, whom we met in chapter 12, noticed that plant elements can be easily identified and categorized. To form flowers on a stem, for example, leaves are sequentially transformed into bracts, sepals,

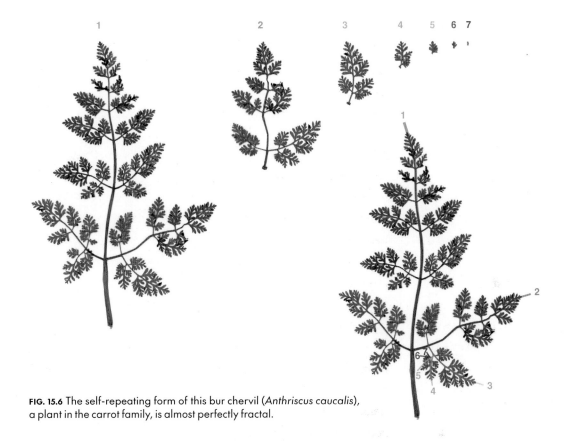

**FIG. 15.6** The self-repeating form of this bur chervil (*Anthriscus caucalis*), a plant in the carrot family, is almost perfectly fractal.

**FIG. 15.7** Here, Stéphane has reversed the order of the previous image and made all the elements the same size. This reveals a nearly perfect series of complexification, with side fingers forming on fingers, similar to the abstract figure of Koch's snowflake.

petals, and stamens. In this way, a plant can be described using a code indicating which element is connected to which, in a linked order.

Lindenmayer then realized that the growth of a plant can also be coded as simple transformations. This led him to create the formal mathematical grammar that many mathematicians and computer scientists have adopted, L-systems. These make it very easy to program the shape of plants—and, more than the shape itself, the plant's actual development. The repetition of the simple evolution of a shape is, in fact, the dynamical way of producing a fractal structure. This is why so many plants have a clear self-similar structure, the most striking example being Romanesco broccoli, the very plant (shown here in figs. 15.1, 15.8, and 15.9) that first drew Stéphane to phyllotaxis.

From the time of Lindenmayer's coding process in the 1960s, computer scientists have worked to make plant pictures increasingly realistic. (Recall Przemysław Prusinkiewicz's sunflower from chapter 12.) Beyond creating fabulously detailed images, the developmental processes they use are based on the dynamics of growth. A growth program can now be made so realistic that computer scientists call it a "numerical seed," which will grow according to its (numerical) environment. With development encoded, there has been a shift toward understanding the process of biological development itself. This shift is similar to the one seen historically in the work of Hofmeister, who moved from describing a static plant structure to understanding its development over time.

Undeniably, L-systems are powerful, capable of describing growth in plants and their possible fractal shapes. Yet scientists are still trying to unlock the basic set of rules that instruct a plant to put a leaf here, a stem there, and so on. Simply trying to state the rule for where a plant places its next primordium has generated a lengthy discussion in this book.

To see how self-similarity works in the Romanesco broccoli in figure 15.8, first cut out one "cone." Draw phyllotaxis spirals on it in both directions (using Fibonacci numbers, of course). At each place where the spirals cross, place the same cone you started with. Then repeat the process. At bottom right, the smallest cone is already at step 3, as we see a cone with side cones bearing small side cones. The full Romanesco broccoli reveals this process iterated seven times.[11]

**FIG. 15.8** The fractal process shown for the bur chervil leaves works equally well with the Romanesco broccoli we saw at the beginning of this chapter: we can cut successively smaller heads that are nearly identical to the larger head.

**FIG. 15.9** Similarly, if we make all the pieces the same size, we see the successive iteration of the same complexification in reverse order.

### Is the Van Iterson Diagram Fractal?

Before exiting this chapter, we pause to note that Chris and Stéphane had one more fractal debate. This time it concerned the Van Iterson diagram. Is the image that lies at the heart of this book a fractal, or does it fail the test? (For more on the debate, see the appendix.) Hopefully, they're not being self-repeating here!

**Chris says no! (as a mathematician):** It's tempting to say that the tree figure supporting the Van Iterson diagram is a fractal: it exhibits beautiful self-similarity that propagates all the way down. But take any point on the tree and arm yourself with a powerful enough microscope. What you'll see is, for most points, one single smooth arc (a branching point will have three arcs stemming from it). That's a far cry from the unending complexity of nooks and crannies you would observe *at any magnification* near an arbitrary point on the Koch curve. Because of this, the Van Iterson diagram has dimension 1, in any reasonable sense of dimension possible.

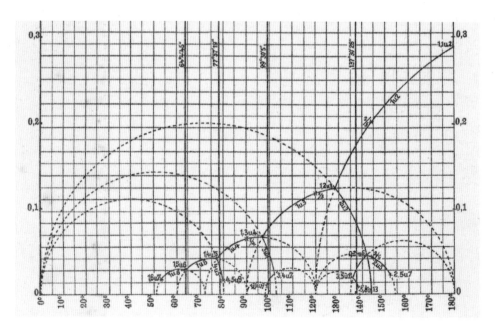

**FIG. 15.10** The branching Van Iterson diagram, once again, for reference.

In other words, if we stick to Mandelbrot's own definition of fractal, the Van Iterson diagram is *not* fractal. If you decide to choose a looser notion of fractal that rests only on self-similarity, you will have to accept that a line segment is fractal. After all, a line segment is also made of smaller segments pieced together that look like the original one, but at another scale. This is a conundrum that Mandelbrot himself seemed to have battled with.

**Stéphane says possibly yes! (as a physicist)**: Let's look again at the self-similarity of the Van Iterson diagram, which reveals a richness in itself. The Van Iterson tree falls into the same category as real plants: at some scale it is really like a fractal. It is constructed step by step, applying a simple transformation to slowly complexify the structure. But it does retain some smoothness. Therefore, to mathematicians, plants and Van Iterson's tree are not truly fractal.

Still, it's striking how close plants can come to the pure mathematical fractal ideal. Plants must live in the real world, after all, and therefore perhaps we should adapt our definition. In their real-life existence on Earth, plants are limited not only in their number of iterations but also in their local size. Essentially, this is because of the size of a typical plant cell.

We see this limitation expressed in the smallest possible leaf finger in a fern, for instance, or in a primordium bump in Romanesco broccoli. Even if the broccoli shows seven iterations—which is already a large number—locally the process stops once the structure becomes too thin. We can also see this in the bur chervil: the lower branches, which are bigger, reiterate and add more indentations. But the middle and upper fingers, which are smaller, do not.

Taking a step back, we see this discussion running throughout this book. In plants, we can see nearly perfect mathematical objects. This is very inspiring. Should this beautiful thing be discarded because plants are not truly perfect, in the mathematical sense? Couldn't the imperfections be overlooked or blamed on "mere" reality, compared to the purity of mathematics?

Mathematicians often succumb to the temptation of going "to infinity"—infinitely many times, infinitely thin, infinitely precise—creating barely imaginable and necessarily unreal objects. Instead, shouldn't we look to the true meaning of reality's limitations, and to the beauty of plants' real constructions, beautiful even in their limitations?

Perhaps the mathematician Kenneth Falconer's more inclusive approach to fractals could bring Chris and Stéphane into harmony:

My personal feeling is that the definition of a "fractal" should be regarded in the same way as a biologist regards the definition of "life." There is no hard and fast definition, but just a list of properties characteristic of a living thing, such as the ability to reproduce or to move or to exist to some extent independently of the environment. Most living things have most of the characteristics on the list, though there are living objects that are exceptions to each of them. In the same way, it seems best to regard a fractal as a set that has [certain] properties . . . , rather than to look for a precise definition which will almost certainly exclude some interesting cases.[12]

Having explored this fractious issue, we will move on to another expression of geometry in plants, this time in the leaves themselves.

CHAPTER 16

# Leaf Bud *Kirigami*

In this chapter, Stéphane describes research conducted in his lab that sheds light on how leaves are folded to fit inside buds.

For a growing plant, the packing problem is not limited just to new primordia or dividing cells. A plant must find a packing solution whenever space is tight—meaning quite often, as you can imagine. One interesting example involves the tight space inside a leaf bud, which turns out to directly impact the shape of a plant's leaves. Although Fibonacci numbers do not play a role here, leaf folding reveals another pleasing aspect of plant geometry, one we can replicate in simple form using the Japanese art of *kirigami*.

Back in 2005, Etienne Couturier came to my lab in Paris for training. As a student, he already stood out for his rare ability to observe and question simple things. The first assignment I gave him was not related to leaves, but instead to the structure within a ball of crumpled paper. If you don't happen to have a micro-MRI machine handy, here's the best place to start: take a sheet of paper that's white on one side and black on the other, for contrast. Crumple it up, wrap it tightly with tape to prevent any movement, and then slice it through the middle using a sharp knife.

What we found was that after the initial crumpling, adding more pressure created a surprisingly regular structure folded mainly in one direction. The reason is that when the paper ball resists, it finds a direction of weakness—and

**FIG. 16.1** A crumpled ball of black-and-white paper, taped and cut in half.

it will fold globally along this direction, creating a structure inside that is invisible from the outside.

While I found this observation interesting, Etienne wasn't satisfied. Next, we decided to explore the structure of crumpled leaves within a red cabbage. He discovered that if you cut a red cabbage just above its main stem tip (the apical meristem so often mentioned in this book), then the cabbage's regular phyllotaxis structure appears. You can observe, especially around the leaf veins at the center, the two spirals winding in opposite directions—with, of course, Fibonacci numbers of spirals.

In fact, a cabbage is actually an enormous bud that hasn't opened. Its leaves remain tightly curled around each other in a dense and compact form. As the newer leaves keep growing, they are constrained by the older leaves around them. As a consequence, they crumple. Etienne found that the leaves actually do this in a rather orderly way, with consecutive pairs of leaves crumpling together within the regular Fibonacci phyllotaxis framework. In figure 16.2 at right, the pairs of colored leaves alternate with the pairs of uncolored leaves, highlighting the cabbage's two counterclockwise spirals.

**FIG. 16.2** This cabbage has been cut just above the apical meristem, its white core. At right, coloring the cabbage leaves following one contact spiral shows how pairs of leaves in contact crumple together, growing in a tight space. Do you see two spirals running in the same direction?

After his cabbage success, Etienne came back to the lab one morning and told me, "You know, leaves inside buds are also folded." I replied, "Not true, show me." We headed out to the nearby Paris garden where Etienne had observed this. I had to admit that Etienne was right. We also realized that the leaves were not randomly crumpled inside the buds, but instead somewhat regularly folded. In many plants, such as sycamore maples, one bud actually contained several leaves. Not only did the leaves inside the bud display regular folds, but also the edge of each leaf appeared to be cleanly sliced off. In fact, this neat edge formed where the folded leaves simply could not expand any farther. We realized that this limitation directly impacts the shape of the leaf when it unfolds.

Etienne was hired as a PhD student to tackle this subject. He had to figure out how to describe the folded leaves scientifically, using numbers and graphs. Along the way, he also figured out how to approach the problem using folded paper and scissors—a variation on origami known in Japan as *kirigami* (*kiri* meaning "cut").

In the course of his research, Etienne made several discoveries. He found that the folds with their backs toward the outside follow the successive main

**FIG. 16.3** This close-up of the previous cabbage shows the leaves expanding laterally, starting from the circular primordium at the center and then crumpling. Green lines indicate the three counter-clockwise spirals of adjacent leaves in this (2, 3) configuration.

**FIG. 16.4** A slightly magnified bud of a sycamore maple (*Acer pseudoplatanus*) starting to pop open, sliced crosswise near the top. Four tightly packed leaves are seen here: two very compressed ones toward the center, and two expanding ones at top and bottom.

**FIG. 16.5** These tiny leaves, just emerging from the buds, reveal how they were folded along their veins. The top leaf is from a black currant (*Ribes nigrum*) and the bottom from a Norway maple (*Acer platanoides*). Do you see the sharp folds on the vein side and the more rounded folds on the other side?

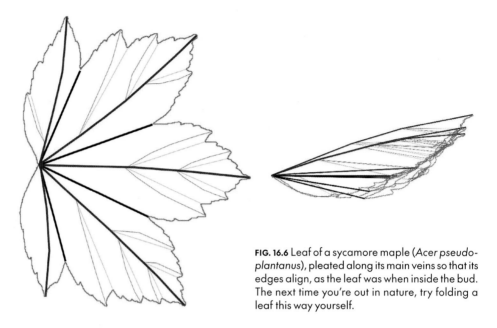

**FIG. 16.6** Leaf of a sycamore maple (*Acer pseudoplantanus*), pleated along its main veins so that its edges align, as the leaf was when inside the bud. The next time you're out in nature, try folding a leaf this way yourself.

veins of the leaf. Meanwhile, the folds going the other direction (with their backs toward the inside) were more rounded. These opposite folds passively allowed the lamina, or flat surface of the leaf, to turn back the other way. You may have noticed that the veins are much larger on the outside (bottom side) of a leaf, a structure that turns the laminae on their side and creates this succession of back-and-forth folds.[1]

When spring comes, the leaf bursts from its bud and unfolds. Usually, it curves exaggeratedly in the opposite direction, eventually settling into a fairly flat state. (This observation, also first met by my "Not true, show me," is still inspiring research in my lab.) Using that information, we could take a mature leaf, look at the main radial veins, and then refold the leaf to see how it fits inside the bud. Even using mature leaves, we could re-create the folds quite convincingly.[2]

Armed with this knowledge, you can easily make accurate *kirigami* leaves out of paper. You simply fold the paper according to a few rules and then cut the edge of the leaf, marking the place where the leaf could grow no larger inside the bud. Unfold the paper and voilà! (See the activity at the end of this chapter.)

Are there broader applications? Definitely. For one, this work has shed light on the connection that paleontologists have discovered between leaf shape and climate. Leaves without many indentations, they have observed, generally occur in warm tropical climates. Leaves with a moderate number of indentations are typically found in temperate climates. And finally, leaves with many indentations—think sugar maple—are usually found in cold climates. This pattern has helped paleontologists make educated guesses about past climate conditions from the fossilized leaves they study. As to the riddle of why, Etienne supplied the answer: in cold climates, leaves waiting to open through long winters have more time to develop inside the bud, creating more folds and consequently more indentations. In warm climates, by contrast, the leaves pop out quickly, before they can develop complexities. This explanation is, however, still debated among botanists.

Another conclusion we can draw from Etienne's work is that there are still many blanks on the map of scientific knowledge. These spots remain empty because no one knows they even exist. Sometimes it takes just one observation and a questioning mind to discover an entire continent that has not been explored, even now. Who would think that a technique as simple as paper folding could solve biological mysteries, in an era when many biologists conduct research at molecular scales?

## A Nod to History

During his investigations into leaf folding, Etienne wondered whether others had come before him. He found that two scientists from Japan—the country of origami—had seen that leaves could be folded along the vein patterns. But their reconstructions of the folded leaf (similar to fig. 16.6) were imprecise, as they did not see the link between the edge of the leaf and the space limitation inside the bud.

Etienne also found a forerunner in the work of eighteenth-century French botanist Michel Adanson, who proposed a new way of classifying plants. At the time, the leading system of classification was that of Carl Linnaeus, who

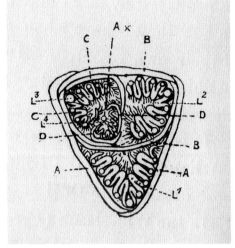

**FIG. 16.7** Michel Adanson, who studied "prefoliation."

**FIG. 16.8** John Lubbock published this drawing in 1899 of a sliced bud of an alder (*Alnus glutinosa*), showing the folded leaves packed inside. His drawing resembles the bud of the sycamore maple shown in figure 16.4.

categorized flowering plants by the number of pistils and stamens inside the flowers. But what if the plant was not flowering?

During his scientific expedition to Senegal, Adanson studied the ways that leaves are folded inside buds—one of his many approaches to classifying plant organs. He called this folding "prefoliation." He then proposed several ways that the leaf might pack itself in order to get maximal extension in minimal volume.[3] In the end, Adanson never reached his overambitious goal of classifying all living plants. But his prefoliation system survived and still occasionally appears in botanical books. The British banker-scientist John Lubbock also used prefoliation to describe buds, noting that buds with folded leaves have corresponding palmate leaves. But surprisingly, he did not see the geometric link between the two.

At this juncture we will move on to a wholly new perspective. How do biologists approach the question of why Fibonacci numbers appear in plants? On this journey, Jacques Dumais will be our guide.

# Try Your Hand

## Make a *Kirigami* Maple Leaf

**DIRECTIONS**

Basically, you will be cutting your leaf from a double-sided fan. Each side will have three matching creases.

1. Take a sheet of printer paper. Ideally, use a page with a contrasting color on one side. Here, the green side represents the top side of the leaf, and the white represents the bottom side.
2. Cut 2″ off the top of the paper.
3. Fold the paper in half vertically, bringing the left edge over to meet the right edge. This central crease represents the leaf's main vein, or midrib.
4. Using the top flap, make the second fold to the left, as shown. Two important tips:
   - The base of the fold must meet the bottom point of the central crease. **All folds in this leaf must meet this bottom point.**
   - The left-hand edge should nearly cover the top point of the central crease, leaving only a small triangle exposed.
5. Make the next fold to the right. Make sure the base of the fold meets the bottom point. With this fold, you have made the largest leaflet.
6. Make two more folds as shown, alternating left and right, and always meeting at the bottom point. Don't worry about making the distances between the creases exactly the same, as real leaves are not perfectly symmetric.
7. Flip over your page. Now make four alternating folds on this side, this time starting with a fold toward the right. Try to match the folds on the first side.
8. Now it's time to cut your leaf. First, flip the paper over to the original side.
9. Next, measure a point 2″ down from the top, at the center crease. This will be your starting point. Draw a smooth curved line from this point down to the base of the leaf, slightly to the right of the bottom point. Two important tips:

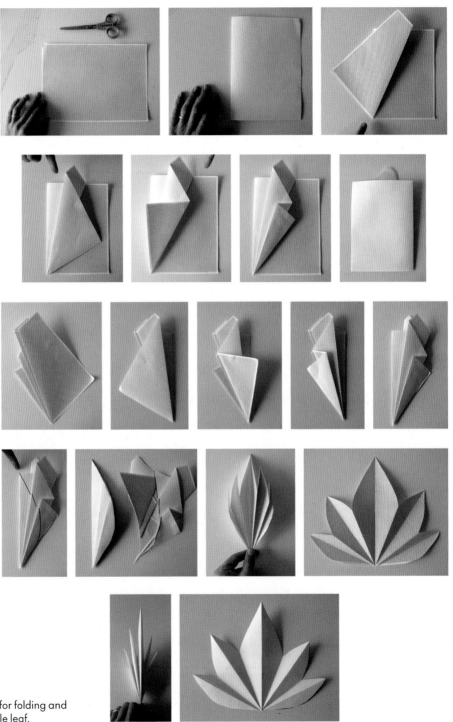

**FIG. 16.9** Steps for folding and cutting a maple leaf.

- ◉ Do not cut straight through the bottom point, but instead slightly to the right of it. (If you cut through the bottom point, your leaf will have holes at the bottom.)
- ◉ Draw your cutting line **below the tip of the last (lowest) crease**— otherwise your bottom leaflet will come out with strange angles.

10. Cut along your line and carefully open your leaf. Experiment with different cutting lines to see how they affect the leaf's contours.

How does this paper leaf compare to a real one? As mentioned above, the central crease represents the leaf's main vein, or midrib. Each subsequent fold represents a lateral vein. In a palmate leaf like this one, all the large veins meet at the bottom point, where the leaf will join the stem. The curved cutting line represents the edge of the leaf, where growth was limited inside the bud by the other leaves.

Making a good leaf takes a little practice. If you learn to do it quickly and casually, using any piece of paper on hand, this is as good as a magic trick!

**FIG. 16.10** Many other leaf-folding variations are possible, some very intricate and others simply torn by hand.

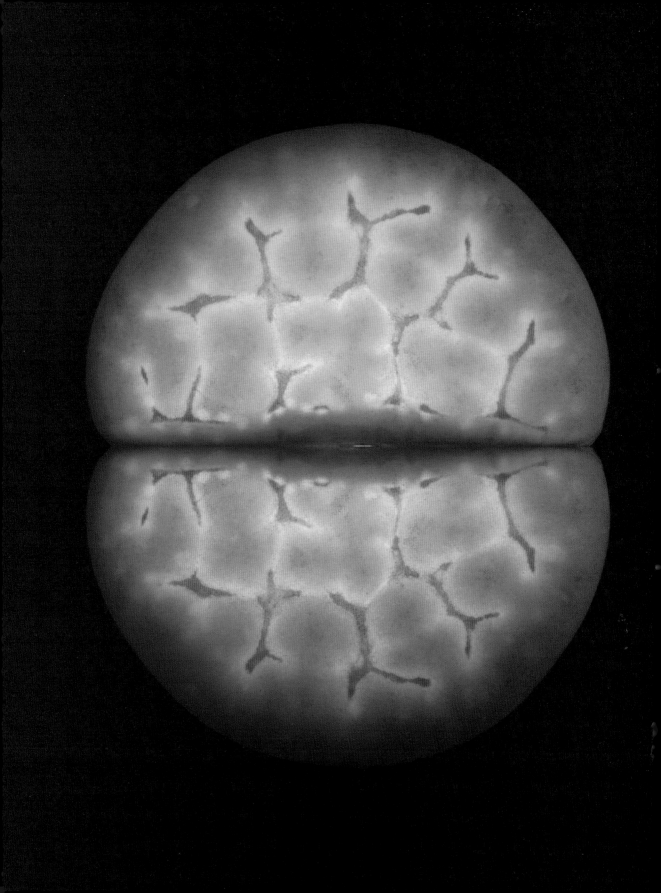

PART V

# What Do Biologists Think?

# CHAPTER 17

# The Hormone That Makes Spirals

What
more
to our
spiral tale?
Well, biology!
Animals aren't the only ones
with growth hormones. Plants have them too, a key one: auxin.

Over the course of its long history, phyllotaxis research has pitted two groups of scientists against each other: the empiricists versus the idealists. For an empiricist, the essence of phyllotaxis lies in what can be inferred from direct observation. For an idealist, phyllotaxis exists above all as an abstract concept in the human mind, with plants serving as mere approximations.

While empiricists revel in the diversity of patterns seen in nature, idealists tend to gloss over the details in search of universal truths. A survey of the nearly three centuries of work on phyllotaxis paints a dramatic scene in which these two worldviews often clash. The harshest criticisms have been aimed at the idealists whose "spiral theory" we described earlier. Julius von Sachs, a prominent nineteenth-century German botanist, saw their approach to plants as "a

FIG. 17.1 Julius von Sachs gazing at a pinecone through the eyes of an empiricist.

sort of geometrical and arithmetical playing with ideas."[1] The work of Schimper and Braun, he argued, was not merely inaccurate but actually stood "in direct opposition to scientific investigation."[2]

Up to this point, our discussion has focused mainly on the mathematical regularity of observed patterns, seen from the perspective of the idealists. Now it's time to give voice to their critics, and to introduce biologists' long-standing efforts to ground phyllotaxis in concrete cellular processes. For this empirical analysis, the natural starting point is once again Hofmeister's rule: the new primordium develops where there is the most available space. As we have seen, idealists accept this rule at face value and use it to justify their geometric models—stacking disks closely together to generate patterns. Empiricists have taken a different approach, however: they have sought to explain how close-packing is achieved through biological processes in the meristem. In other words, how do plants "find" the site with the most available space on the meristem?

Before looking more closely at meristems, however, let's consider another example of biological close-packing whose control mechanism will later serve as our template: the self-assembly of proteins at the molecular scale.

### The Phyllotaxis of Protein Crystals

Of all the biological processes in which close-packing occurs, the one most similar to disk stacking involves the addition of proteins to form biological polymers. These patterns are more readily explained than most, because of the definite size of proteins and the absence of secondary growth processes that would distort regular arrangement.

 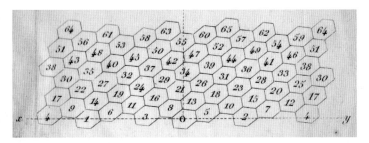

**FIG. 17.2** At left, microscopic viral proteins form a regular lattice in the shape of a hollow cylinder. At right, a cylindrical lattice drawn by the Bravais brothers, who would have been thrilled to know about these microscopic structures.

When you glance at the cylindrical protein structure in figure 17.2, you might think you're looking at an odd pineapple or the trunk of an alien palm tree. In reality, the largest of these structures rarely exceed 1/1,000 the width of a human hair—and they are accessible to us only through the use of powerful electron microscopes. They exist in a physical realm far from any plant pattern we've seen so far. At this scale, patterns are dictated by the fundamental symmetries of matter itself.

In crystal-like fashion, the proteins come together to build orderly cylindrical structures out of tiny "bricks." From an amorphous stew of proteins within the cell, these structures must somehow fall into place, finding precisely the right position and orientation to propagate the crystal. Contrast this with phyllotaxis patterns in plants, whose orderly arrangements emerge from amorphous masses of whole cells. But let's not get ahead of ourselves. How exactly are the protein crystals assembled?

Two simple steps are involved. First, the protein units must be brought together. At the molecular level, this is easily achieved, given that thermal excitation implies a constant chaotic motion of molecules. If enough protein units are present, they will invariably come into contact with each other.

To visualize the second step, imagine the protein units being held in place by tiny springs. When units are brought close together in a favorable alignment, short-range attractive forces keep them in a definite configuration. If the distance is too great, the microscopic forces combine into a net attractive force. If the distance is too short, the forces combine into a net repulsive force. The outcome of the two-step process is the orderly addition of units at the edge of the protein crystal cylinder.

By the time the Bravais brothers published their article in 1837, the main mathematical tools needed to characterize phyllotaxis patterns were already known. Certainly when Van Iterson's book was published in 1907, a very detailed quantitative account of these patterns was available. But these scientists had no knowledge of protein crystals—which would not be discovered until the advent of electron microscopes and X-ray crystallography several decades later. In 1973, the American biologist Ralph O. Erickson showed that tubular protein assemblies are the perfect embodiment of the phyllotaxis theories developed at the turn of the twentieth century. Erickson even proposed a bifurcation diagram for the cylindrical lattices that follows that of Van Iterson very closely.

Protein crystals are remarkable for another reason: perhaps for the first time, they gave both empiricists and idealists a system that explained the patterns observed in nature. No doubt, a satisfactory conclusion has been reached

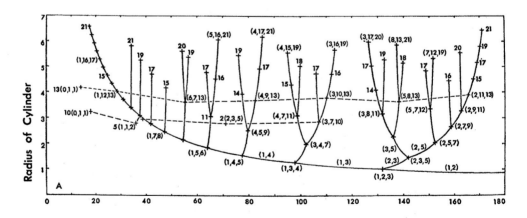

FIG. 17.3 Ralph Erickson's lattice diagram for protein crystals. Here the Fibonacci branch (do you see it?) does nothing special. Cylindrical crystals with a fixed number of orthostichies (here, 10 and 13) are linked by dashed lines, echoing Stéphane's question in chapter 13: How do plants choose among the modes with 13 orthostichies?

thanks to the brick-like role of the protein units and the relative simplicity of their mechanical interactions.

So now, have we also arrived at a clear understanding of how phyllotaxis patterns come about in plants? Unfortunately, not quite. When observed under a microscope, most plant organs are not in close contact with each other at the time of their formation. Rather, it is as if an invisible force allows young organs to form neither too close to nor too far from the organs already in place. This invisible force turns out to be a powerful plant growth hormone called auxin.

## Auxin and Pattern Formation

As you may recall, Alan Turing showed that reaction and diffusion can give rise to regular patterns on a surface, through the actions of an activator and an inhibitor (see chapter 11). Currently, the biological mechanism that seems to best explain phyllotaxis is similar to Turing's theory, with auxin playing the role of activator and active transport contributing to the diffusion of molecules.

How exactly does this work? The developing primordia deplete auxin in their surroundings, depriving cells of an essential ingredient to initiate a new primordium. *As a result, each new primordium creates a zone of inhibition around itself.* Although auxin could in theory promote formation of primordia in the meristem center, the cells of the central zone are not yet able to respond to the growth hormone. And so, new primordia form just outside the central zone, in the widest gaps between the older primordia. These are the sites where the cells are both competent to form primordia and also exposed to the highest auxin levels.

Unlike for the mechanical packing of proteins, however, it is not immediately clear that

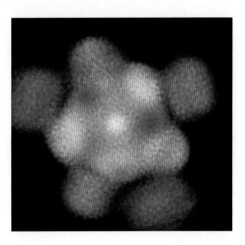

**FIG. 17.4** Here, cells inhibited by lack of auxin light up in bright yellow. No primordia can form in the pink circle, with or without auxin. The darker in-between areas are where new primordia can form. Can you guess the next best place?

inhibition around new primordia can lead to regular phyllotaxis patterns. Auxin inhibition explains only the primordia interactions. In the early 1900s, the Dutch botanist Johannes Cornelis Schoute made one of the first attempts to understand the "geometry" of inhibition in guiding the placement of plant organs. Prior to Schoute, scientists had drawn phyllotaxis patterns using circles (or other regular shapes) to represent the contour of the organs on the meristem. Schoute took a different approach, instead drawing zones of inhibition that overlap.

In this way, Schoute's drawings offer a degree of abstraction not seen in prior research. Interestingly, the zones of inhibition drawn by Schoute remained invisible to experimentalists for nearly a century. Only recently have they been observed in living meristems, using fluorescent reporter proteins, as shown in figure 17.4. These inhibition zones are regulated, more or less directly, by the concentration of auxin in the cells.

Figure 17.5 shows Schoute's illustrations of two possible models explaining primordia growth, redrawn here for clarity. At left is Schoute's concept of inhibitory fields. The overlapping circles represent the zones of inhibition that arise from the presence of primordia $u$ and $v$ on the meristem surface. A new primor-

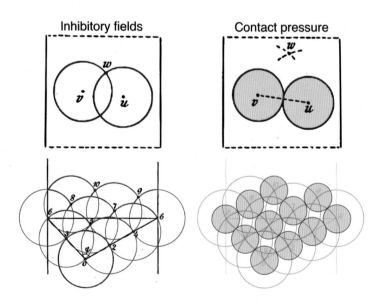

**FIG. 17.5** At left, Johannes Schoute illustrates his concept of overlapping zones of growth inhibition (blue circles). It results in the same phyllotaxis patterns as the older model of contact pressure, leading to the stacked red disks shown at right. Do you see a fundamental difference?

dium ($w$) is placed in the "first available space" above the zones of inhibition. For this mechanism to work, the primordia do not need to be in physical contact, and in fact they could occupy a very small region of the inhibitory circle.

In the right-hand illustration, the tangent circles represent the familiar contact model: the primordia are in physical contact with each other on the meristem surface. A new primordium ($w$) is placed as low as possible on the meristem without overlapping with preexisting primordia. Note that Schoute's inhibitory fields model predicts the same phyllotaxis patterns as the older disk stacking model used in nearly all other mathematical studies of phyllotaxis. This would explain why the two models have coexisted for so long.

While a purely mechanical explanation of phyllotaxis would be untenable today, some scientists are reconsidering whether mechanical pressure could still be important in meristems. One intriguing possibility envisions it as a feedback mechanism, maintaining the integrity of the meristem as it grows and initiates new organs. In a sense, Schwendener's contact pressure theory lives on through these studies. Mechanics also plays an unlikely role in the division of plant cells, this time with the key mechanical principle being not contact pressure but surface tension. To further explore these connections, our next chapter will look at cell division and soap bubbles.

CHAPTER 18

# A Cell Division Discovery via Soap Bubbles

Think
soap
bubbles.
That's what Jacques
did, when puzzling out
how cell division works in plants.
Oddly enough, the answer led to a real puzzle.

The following chapter offers a second tale from biology, this time shedding light on cell division in growing plants. Although cell division is not part of phyllotaxis studies per se, Jacques's research dovetails nicely, offering a parallel example of geometric and mathematically predictable plant patterns—understood through their dynamics. Most strikingly, Jacques's research culminated in a cell division diagram whose structure closely parallels that of the Van Iterson diagram. Essentially, Jacques's discovery of his cell division "tangrams" is, he said, about "how science advances in very strange ways." As is only fitting for a book about spirals, his work serves as a perfect example of how science itself can be a spiral.[1] "Scientific progress does not follow a straight line from question to answer, or from problem to solution," Jacques observed. "Rather it spirals inward infinitely and ever closer to the answer."

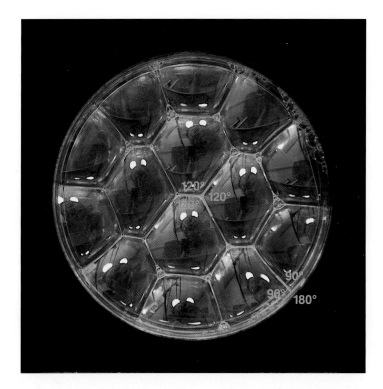

**FIG. 18.1** Soap bubbles confined to a circular dish. For Jacques, bubbles would help shed light on cell division in plants.

Although Jacques had been fascinated by phyllotaxis since his undergraduate days at Sherbrooke University in Québec, his research took him in other directions. Fast-forward to 2007, when, after obtaining his PhD, he was working as a biology professor at Harvard. A colleague asked him to write an article about the role of mechanics in plant pattern formation. At the same time, he recalled, "I had a high school intern that I needed to keep busy, so I said, 'We'll use soap bubbles.'"

Jacques and his intern went back to a neat formulation of cell division set down more than a century ago by Léo Errera, a plant scientist in Belgium. In one of those leaps that finds simple underlying rules in complex behaviors, Errera noticed that the walls of plant cells behave like soap bubbles. No doubt, Errera was influenced by another Belgian scientist, Joseph Plateau, a physicist who in the mid-nineteenth century had been experimenting with the laws governing

**FIG. 18.2** Portrait of Léo Errera.

the equilibrium of soap bubbles. Plateau was, in fact, observing the effects of intermolecular forces. He concluded that soap bubbles ought to adopt shapes that minimize their surface area.

To understand the connection with plants, it's helpful here to review a bit of basic biology: unlike animal cells, plant cells are surrounded by a rigid wall. These walls help plant cells hold their form, since plants have no other exoskeletons or bones. Scientists have long been interested in how plants "know" where and how to set down a new cell wall as the growing plant forms new cells. Errera came up with a possible explanation involving soap bubbles. His rule from 1886 states: "The cell [wall], at the time of its formation, adopts the geometry that a soap film would take under the same conditions."[2] The work of Plateau, Jacques said, "inspired him to formulate his division rule. Dividing plant cells are like soap bubbles. It's an analogy."

Heading into this project, Jacques was vaguely familiar with Errera's rule, but he knew the concept better from the version set forth by D'Arcy Wentworth Thompson in his famous book *On Growth and Form*.

### Bubbles in the Lab

Thompson was a Scottish biologist, mathematician, and classics scholar whose work has itself been a classic since it was published in 1917, offering observations on a staggering range of topics, from zebra stripes to sand dune shapes to phyllotaxis. While *On Growth and Form* was a serious work of science containing no shortage of equations, Thompson was also something of a popularizer. "He had a big book that lots of people wanted to read," Jacques said, "and he liked to make little drawings and predictions." Among the formulations that Thompson

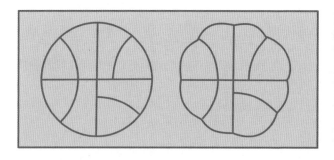 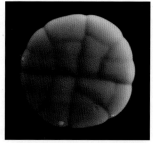

**FIGS. 18.3** At left, Thompson predicted the division of a cell into eight octants. His drawings are strikingly similar to the SEM image of a digestive trichome of a Venus flytrap (*Dionaea muscipula*) at right. Which octant seems to be flipped with respect to the others?

took on was Errera's rule. "The genius of the original rule," Jacques said, "is that Errera avoided some pitfalls by not giving you anything beyond the soap bubble analogy." Thompson's variation, however, provided something concrete that could be measured.

In restating Errera's rule in a chapter on cells and tissues, Thompson took the analogy one step further. He stated: "The incipient partition-wall of a dividing cell tends to be such that *its area is the least possible by which the given space-content can be enclosed.*"[3] Jacques calls this "the principle of minimal area" and restates it a bit more simply: when a cell divides, the cell wall adopts the configuration of least area that encloses a fixed cell volume.

Jacques and his intern took on the challenge of seeing whether Thompson's drawings could be produced using real bubbles. They mixed regular dish soap with a little glycerin for stability, and then they manipulated the bubbles in a petri dish. They used two syringes to blow conjoined bubbles as a simple way to emulate cell division. Inside the circular dish, one large bubble was divided into two neat halves, then into four equal quarters, and eventually into smaller divisions with subtle but predictable curves. Each time, they carefully replaced the fluid soap films with identically shaped plastic walls so as to mimic the rigid walls of plant cells. As they studied their bubbles, Jacques said, "we took pretty pictures, and we even made movies."

**FIG. 18.4** At far left, cell divisions in a fully developed glandular trichome from a Venus flytrap leaf. Starting second from left are soap bubbles in a circular dish, showing sequential subdivision. Do you see the two types of octant cells?

Next, they compared the bubbles to actual plant cells. They turned to the Venus flytrap, a carnivorous plant that grows trichomes, specialized hairlike structures that, in this case, secrete enzymes that can digest prey.[4] The trichomes' cells form "beautiful predictable patterns," Jacques said. Even better, they fit perfectly the configurations predicted by Thompson on the basis of the least-area division rule.

So now they had successfully tied Errera's rule to actual bubbles and actual plant cells. But Jacques was also interested in determining how cells would divide if they were asymmetric or irregularly shaped. He turned to a paper published in 1914 by Norbert Wiener, a child prodigy who completed his PhD at Harvard at the age of 19, writing a dissertation on mathematical logic. (At MIT, Wiener would go on to found the field of cybernetics, the science of communication and automatic control in machines.)

The problem Wiener laid out (paraphrased here for clarity) was this:

Let us suppose a farmer wants to divide an irregular field equally between his two children, using as short a fence as possible. What shape should the fence be?[5]

The solution to this problem shed light on Jacques's cell division work—even though Wiener never referenced Errera and predated Thompson's book. Basically, it's a two-dimensional problem, Jacques explained, in which you're looking for ways to minimize length inside a weird shape. "That," he said, "is exactly the same problem we had to solve."

Wiener's solution does not involve a one-size-fits-all equation but instead lays out certain geometric properties. One part must be an arc of a circle: this is because the geometric figure with the shortest perimeter for a given area is a circle. "Wiener shows that the solution should be an arc of a circle meeting the edge of the field at ninety degrees," Jacques said. "The actual position depends on the shape of the field." That was exactly the same conclusion that Jacques and his colleagues reached when looking at the cells of the Venus flytrap. And that's what Jacques wrote about in his review article.

But the truth was, this approach didn't tell the whole story. "Most of the cells were what Thompson predicted, in terms of minimizing area. But oh—look at that! There are some cells that don't fit the pattern," Jacques recalled. While the soap bubble model usually worked, it wasn't universal, Jacques explained. "When Thompson bastardized Errera, or when Wiener tried to state something more formal, they both said: I'll look for the shortest wall possible. What we saw in the Venus flytrap gland cells was that *most* divisions were minimal, but there were other divisions that were *not* minimal." Usually these unexpected divisions corresponded to "local" solutions of Wiener's problem. More specifically, the walls were shorter than other, similarly placed walls—but better configurations were also available beyond this small neighborhood. "We saw that in both plants and soap bubbles," Jacques said, "the divisions were not confined to a single solution."

## A New Cell Division Rule

Up to this point, Jacques's discoveries were linked to past research done by others. It took another four years for him to move to the next phase, looking for a rule that would cover a greater range of divisions in plant cells. Once again, he turned to soap. In 2011, he and Sébastien Besson, a French postdoc at Harvard, began blowing lots of bubbles. This time, they used a glass needle to move the walls that separated two bubbles, to see what equilibrium configurations they would take. As Richard Grant wrote in a 2011 article about their work, they observed that "no matter how many different ways they drew the glass needle, the bubble would go back to one of a fairly limited number of shapes."[6]

At the same time, they observed similar behavior in plants. "We looked at the cells of different plants, including algae, zinnias, and the Venus flytrap," Jacques said. "We found that all of these cells—despite being very distant in their evolutionary history—have similar competing solutions for where they divide. It's as if the cells can smell that there's a good solution." Take, for example, a cell shaped like a long rectangle. In this case, there's one transversal division that would clearly be a local minimum. "If there's a big difference between the two solutions for where to divide," Jacques said, "then plants will only choose the shortest one." But if the cell is more circular, then the choice of where to divide will be less obvious, leading to more than one reasonable outcome.

In other words, a cell when it divides has several options. "Instead of being the shortest fence possible to divide the field," Jacques said, "if the cell shape is fairly complex, then there are many local solutions that are possible, and those are seen as well." His paper discusses not cut-and-dried outcomes but instead *probabilities*: how likely it is that the cell will place its wall at the lower or higher minima.

The result was a new rule for cell division in plants.[7] It states the probabilities governing where cells will divide over the full range of plant life, from flowering plants to pine trees to ferns—and even some algae. Put simply, plant cells choose their division plane among the local area minima dictated by their shape, favoring those minima that come closest to the global minimum.[8] Beyond the simple soap bubble analogy, the new rule applies more broadly.

As for Thompson, he had gone astray. "In simplifying Errara's rule, Thompson threw out the baby with the bathwater," Jacques said. "This became clear when Thompson started making concrete predictions. He assumed each cell shape would be associated with a single division, the shortest one possible. But in fact Errera was careful to leave open the possibility of a cell adopting more than one division plane."[9] Starting from Thompson's failed attempt, Jacques said, "we were able to capture the full breadth of Errera's division rule—but this time in concrete, operational terms." One cell shape, they showed, can lead to more than one cell division plane.

## Tangrams

But the story was not over yet. In fact, it took a very interesting turn. Back in 2011, Jacques had shown his cell division paper to a Harvard colleague, computer science professor Radhika Nagpal, who uses models of bee swarms and termite colonies to engineer new types of collective systems. "She noticed that our data were curiously distributed when graphed," Jacques said. "The data looked like a trumpet, but we never really questioned why. So we started thinking about it."

One possibility involved using the dynamical systems we've seen earlier in this book, which describe the evolution of a complex system over time. "We laid down the curve of a dynamical system onto this cloud of points, and it fit perfectly," Jacques said. "All this because a colleague was perceptive enough to ask, 'Why does your cloud of data look like this?'"

Nagpal's observation spurred Jacques to take a new approach. He realized that a cell isn't static but is always growing and dividing over time. "Grow, divide. Grow, divide. A plant cell is a dynamical system, and we should be able to find where the division rule takes them when applied over many generations. We should get to an *attractor*."

An attractor, within the field of dynamical systems, is a set of states (numbers, shapes, etc.) that a system tends to converge to, from a wide variety of starting conditions. When Jacques applied his cell division rule over and over again, approaching an infinite number of repetitions, he made a startling discovery: the mother cells and their progeny formed logarithmic spirals just like those described by the early students of phyllotaxis. Not only that, the space of potential cell attractors had a structure reminiscent of the pruned diagram developed for phyllotaxis by Van Iterson.[10]

Now he wondered: Would the theory match the actual plants? To find out, Jacques needed to observe the mother of all cells, the single microscopic apical cell.[11] Happily, the observations he collected matched the dynamical system predictions. During the growth process, the mother cell of the apical meristem sheds new cells, its "daughters." Jacques could see that the shapes of the meristem cells converge to one of the several attractors they found. In mosses and

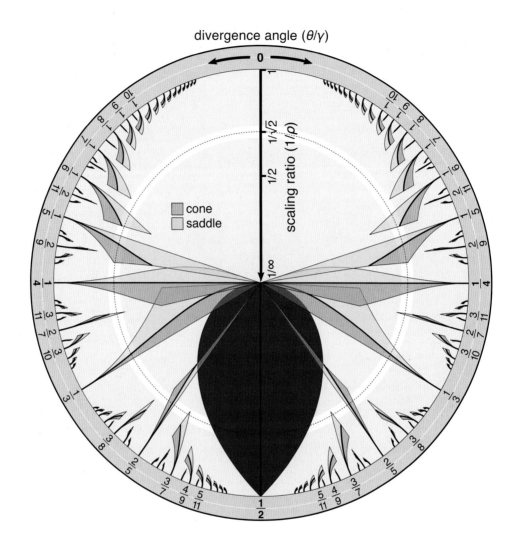

**FIG. 18.5** Van Iterson's diagram, reconfigured to show the solutions of the cell division rule on conical or saddle surfaces. The diagram is shown in polar coordinates. The fractional numbers on the dial represent the divergence angle (in fractions of a turn) and can be interpreted as discussed earlier for phyllotaxis patterns. For example, a divergence angle of 2/5 represents a five-sided cell that must go twice around its circumference as it sheds five consecutive daughter cells. The radial position captures the rate of cellular growth between each division and is akin to the "rise" used to describe phyllotaxis patterns. The left–right symmetry reflects the two possible mirror images for each cellular shape (right-handed spiral vs. left-handed spiral).

 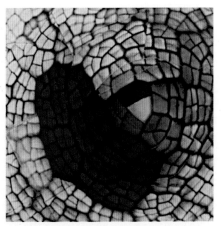

**FIG. 18.6** The cell division attractors. At left, for children, Jacques created a puzzle version showing six common types of cell division in plants. At right, the triangular apical cell of a fern meristem. Can you find it in the puzzle?

ferns, for example, the apical cell is commonly three sided, one of the dominant attractors found in the theoretical analysis.

Jacques calls the predicted cell shapes "tangrams," referring to the Chinese geometric puzzles. Norbert Wiener, in approaching his farm field division problem, had seen that the solution involved arcs of circles meeting the edge of the field at 90 degrees. Jacques's cell division attractors fit these prescriptions quite neatly.

In figure 18.6, note the similarity to the triangular attractor predicted by the cell division rule. In the puzzle version, the triangle appears at top right. It is nothing short of remarkable that for so many plants, the dividing apical cells fit these regular tangrams. And in terms of plants knowing math, this discovery is absolutely elegant.

## The Future

In their ongoing research, Jacques and his students are looking at what happens inside plant cells that could explain the formation of these unique shapes. "A cell is neither a soap bubble nor a farm field," Jacques said, "so we're looking at what goes on inside the cell." One possibility involves microtubules—protein

**FIG. 18.7** Epidermal cells on a spiderwort (*Tradescantia*) leaf. The yellow stoma opens and closes to regulate respiration. Do you see soap bubbles, as Léo Errera did?

filaments throughout the cell that provide structure and transport substances. "It may be," Jacques said, "that the dynamics of these microtubules dictate how cells divide, forcing them into these remarkable shapes."

Now that we have pondered whether plants know math, observing their patterns from many different perspectives, perhaps we might ask the same question about animals.

CHAPTER 19

# A Brief Detour to Animals

Over the course of our long and spiraling botanical journey, you may have wondered whether Fibonacci numbers and parastichy patterns might also have evolved in animals. The answer is somewhat complicated.

The question of whether plant and animal spirals might be linked is a very old one, as seen in the correspondence of none other than Charles Darwin. In 1862, Scottish paleontologist Hugh Falconer sent Darwin an article introducing him to the idea of Fibonacci phyllotaxis.[1] (Darwin joked that the topic would drive him to a miserable death; see below.) These spirals exist not just in plants but also in animals, Falconer believed, citing the shells of mollusks. He told Darwin that therefore there must exist a principle even more powerful than natural selection—"a deeper seated and innate principle"—that would explain these universal spirals. Was Falconer right?

Let's look at some likely possibilities among animals. As we have seen, phyllotaxis patterns appear when a plant is built sequentially, with organs of similar size and shape. From this perspective, it would be silly to argue that the five fingers of the human hand provide evidence of Fibonacci numbers in mammal development. Instead, our search for plantlike patterns in the animal kingdom should focus on structures composed of many nearly identical units. A good place to look is the skin of fish, reptiles, and birds, with their many identical elements. Often, their scales and feather follicles are arranged in a regular hexagonal pattern.

As it turns out, however, there are crucial differences that affect the patterns:

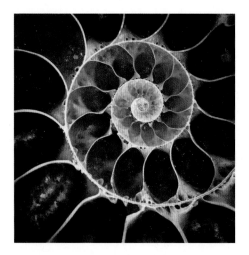 

**FIG. 19.1** Might the spirals in mollusk shells be Fibonacci spirals, converging to the golden mean? Here, two mollusks display their curves, at left a sliced fossil ammonite and at right a lightning whelk (*Sinistrofulgur perversum*, named after its left-handed spiral).

**FIG. 19.2** Might fish scales have Fibonacci spirals?

- Plants have only a top and bottom imposed by gravity.
- Animals, which are mobile, also generally have a head end and a back end.

- Plants can be understood as a vertical axis, elongating upward.
- Animals are bilaterally symmetric on the outside—they need to turn left or right with equal ease.

- Plants have a single developmental front that follows the growth of the stem upward.
- Animals have at least two developmental fronts that form on each side, independently. They propagate sideways and up or down, starting from the spine or belly.

In animals, these developmental fronts can leave behind a regular Turing pattern in the scales, either as simple bands of scales or as hexagons.[2] Yet along an animal's spinal column, the two sides of the pattern are not always continuous. This presents a stark contrast to plants, whose circular developmental front (as drawn by Turing in fig. 11.6) leads to spiraling parastichies. The "connection problem" in animals can be clearly seen in fish, where the scale patterns meet around the upper fins but do not join (see fig. 19.2).

In some snakes, the disconnect is very clear: a mismatched line along the spine reveals where the two separate sides of the pattern meet (see fig. 19.3). This is not always the case, however. In snakes with fewer scales, the two sides of the pattern may interlock. Then we find perfect hexagons (see fig. 19.4). In these snakes, the pattern is similar to a perfect whorled phyllotaxis pattern in plants, with various numbers of symmetric spirals. But if you count the spirals, do you suppose you will find Fibonacci numbers?[3]

The same disconnection problem can be seen in bird feathers. (For evidence, look at the skin on a chicken's back.) Yet there are examples of interlocking patterns in feathers, just as in snake scales—and in fact this interlocking explains the pattern in a peacock's outspread tail. On its back, the peacock's feathers

**FIG. 19.3** In this Arafura snake (*Acrochordus arafurae*), do you see the line of defects along its spine, revealing where the two sides of its scale pattern meet?

**FIG. 19.4** In this Taiwan burrowing snake (*Achalinus formosanus*), the two sides of the scale pattern are interlocked, forming a single coherent pattern.

**FIG. 19.5** A peacock showing off its spirals. How many parastichies are there?

form a regular hexagonal pattern, with the last scales developing into longer and longer feathers, each terminating in an "eye." When the peacock raises its tail, the same way our skin gets "goosebumps" when we are cold, this symmetric hexagonal pattern is on full display. But could the pattern possibly come full circle, as it does for the interlocking hexagonal snake scales in figure 19.4?

The same pattern interruption between top and bottom can also be observed in a very different case, that of lamprey teeth. Here, the packing of the sharp little teeth leads to a pattern that is not the same all the way around the circle but is merely bilaterally symmetric.

Now, what about invertebrates? The compound eyes of insects (whether or not you'd like to stare into them) present a compelling case. In flies, moths, and ants, the photoreceptor cells called ommatidia display a stunning regularity. Once again, we find the same local developmental fronts as in snake scales and bird feathers. This time, however, the front sweeps across the surface of the eye

**FIG. 19.6** The mouth of a lamprey, with spiraling rows of razor-sharp teeth. Do you see the typical left–right symmetry found in animals? And the back–front asymmetry?

as it develops, "transforming a sea of undifferentiated cells into ordered rows" of ommatidia and filling up the eye field, as biologist Justin Kumar has described the process.[4]

Because the eye's surface is curved, the length of the ommatidia front varies as it sweeps through. As a result, the same geometric effects appear as in plants when the meristem diameter changes—three of these effects are visible in figure 19.7. The first is a shift between hexagonal and square patterning. At bottom left, hexagonal ommatidia in the green band become square where the orange band crosses, before changing back to hexagonal again. (You may recall similar transitions from square to hexagonal in the unrolled pineapple of fig. 10.6).[5] The second transition is an isolated "dislocation," where one row branches into two, as seen in the center image.[6] (Recall the flower spike in fig. 14.11). And finally, the orange pentagonal cells in the right-hand photo are either at the center of a dislocation or else isolated pentagonal cells, necessary when covering a full sphere with hexagons.[7]

Yet while the local geometry of these ommatidia patterns is strikingly similar to that of plants, we find a now-familiar difference: the morphogenetic fronts in a fly's eye don't come full circle. Even if there are indeed parastichy-like rows, they are only local and do not circle around a (stem) axis. The prospect that these rows might follow a Fibonacci-like law is therefore hopeless.

Next, let us return to the case of mollusk shells. We will not mince words here: *there is no relationship between phyllotaxis and spirals in mollusks*. Some mollusks have shells that spiral out from the larval axis as they develop. The process is indeed similar to that of building a spiral using Fibonacci numbers

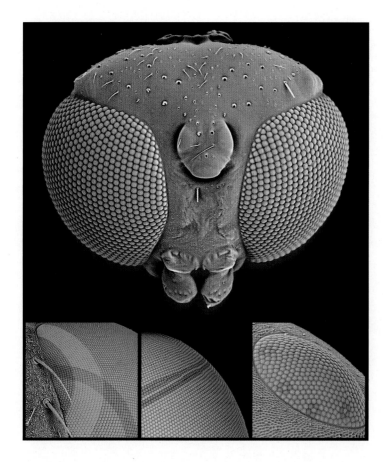

**FIG 19.7** The compound eyes of insects show patterns that are somewhat plantlike. At top, the eyes of a fly. At bottom, three examples of pattern transitions in, from left to right, a housefly, a moth, and an ant.

(as in the chapter 1 activity), but this process has nothing to do with the way that plants build spirals. For a mollusk, the resulting spiral is logarithmic (or very close to logarithmic).[8] If you measure the growth rate of a mollusk shell, you will see that the rate is fixed for each species, and that different mollusk species grow at different rates.[9] Once again, neither Fibonacci nor the golden ratio appears in any of them.[10]

At this point, having concluded that animals cannot form the same Fibonacci spirals as plants, in part because of their bilateral form, we will (after one more short detour) return to the question that launched our entire investigation, the one asked by the title of this book.

### The Role of Natural Selection

We pause here to touch on one last important topic before the book concludes. What about the role of natural selection in Fibonacci phyllotaxis? As mentioned at the outset of this chapter, when Charles Darwin was first introduced to phyllotaxis in the fall of 1862, the phenomenon drove him to distraction. For the next several months, his letters to his fellow scientists kept returning to the topic and whether natural selection could possibly explain it.

The following spring, Darwin's note to Asa Gray, a botanist and friend at Harvard, revealed his state of mind:

> Have you ever formed any theory, why in spire[s] of leaves ... the angles go $\frac{1}{2} \frac{1}{3} \frac{2}{5} \frac{3}{8}$ &c—Why should there not be $\frac{1}{4}$ or $\frac{1}{5}$? It seems to me most marvellous—There must be some explanation. If you have [a] theory, I know it would be too long to explain; but I should like to hear whether you have. My good friend Falconer has been twitting me that these angles go by as fixed a law as that of Gravity & *never vary*.[11]

A month later, Darwin wrote to Gray again, this time in near despair: "If you wish to save me from a miserable death, do tell me why the angles of $\frac{1}{2} \frac{1}{3} \frac{2}{5} \frac{3}{8}$ &c series occur, & no other angles.—It is enough to drive the quietest man mad."[12] Unfortunately, Gray was little help, replying, "I have no notion in the world why the angular divergence should be of *that series* of nos. & not of others."[13] He wrote to Darwin again the following month: "Pray don't run mad over phyllotaxis! I can't save you, I am sure."[14]

This short episode in the life of one of the greatest scientific minds of the nineteenth century once again highlights the awe and obsession inspired by the regular patterns seen in plants. We hope that a reader having come this far in the book could offer Darwin a cogent geometric explanation for why the angles of $\frac{1}{2}, \frac{1}{3}, \frac{2}{5}, \frac{3}{8}$, and so on are those observed in nature. Should you feel underprepared to face the father of evolution himself, take a moment to review chapter 14 on the topic of zigzag fronts.

While the *dynamical* explanation for Fibonacci phyllotaxis is now settled, it was the role of natural selection that kept Darwin up at night. Putting the question very simply: What is the value, if any, of specific phyllotaxis patterns for the survival of a plant? Two main functional explanations have been proposed, one involving the capture of sunlight and the other the optimal packing of plant organs. In the introduction to this book, we debunked the idea that Fibonacci phyllotaxis evolved from plants maximizing sunlight exposure on their leaves.[15] (Recall that the leaves of many plant species grow directly above one another, as in the common whorled phyllotaxis pattern, and also that leaves are capable of moving to catch more light.)

The second possible functional explanation is that Fibonacci spirals might optimize the packing of young plant organs in the meristem and bud, protecting these fragile structures.[16] Since disk stacking already provides a pretty good solution, there was no need for (or room for) another way of growing to evolve. The study of quick disk stacking introduces a mystery, however, as we found that quasi-symmetry tends toward a square state—not the optimal hexagonal packing. One possibility is that there might be some dynamical packing efficiency in Fibonacci phyllotaxis that provides a slight evolutionary advantage, but not enough to forbid the existence of these square modes. Ultimately, the value of Fibonacci phyllotaxis for a plant may well be the next frontier in phyllotaxis research.

**PART VI**

# Conclusion

CHAPTER 20

# Do Plants Know Math?

For most of scientific history, plants were in some sense overlooked. We loved them, we used them, we needed them—but not many scientists felt compelled to analyze their formal structures, which turn out to be surprisingly mathematical. Why didn't we see this before? We hope that this book will lead to more observations, and that our relationship to plants will become closer and even more mutually beneficial.

Over the past three centuries, the subject of phyllotaxis has been like a mythical sea monster: some saw the spirals, but then their work was forgotten, and then someone else saw them, and so on. Progress was incredibly slow, with people unwittingly rediscovering work that had already been done. Each time, there was the same amazement that an impressive mathematical world would be lying unseen in plants. But for a long time, most discoveries were related to the mathematical properties of the Fibonacci sequence—and to the related final limit of its quotients, the golden mean—rather than what happens in real plants. We had to await the stubbornness of Hofmeister and the genius of Turing to begin understanding that these mathematical marvels are built little by little, with simple steps, sometimes with errors and missteps—and not imposed from above by some mathematical deity. In this way, this plant story is most of all a lesson in humility: the abstract patterns we praise so much are not preordained; rather, they emerge from morphogenesis.

On rare occasions (too rare), plants have directly inspired mathematics. Auguste Bravais looked at plants, made an abstract model of them, and then applied this model to seemingly unrelated problems: the structure of crystals and their groups of symmetry. More often, though, the mathematical contributions were missed. When Schimper and Braun made their trees of the possible rational divergence angles, no one made the leap to the structure of rational numbers, later formalized as the "Farey sequence." (Really, we could call it the Farey-Haros-Schimper-Braun sequence.) Similarly, it took a long time for the self-similar structure of the Van Iterson tree to be understood, disseminated, and eventually recognized as a renormalization. Both concepts were independently rediscovered much later. The general lesson we can draw is how much beauty—real mathematical beauty—lies at our feet. We must simply open our eyes. But once we start looking, the journey is endless.

Another lesson is that patterns are not fixed from the start, derived from idealistic rules that early scientists believed to be preordained. Instead, many abstract objects and rules are built little by little. This is the key progression we saw unfolding in the study of phyllotaxis over time: moving from a static idealistic description to a dynamical one that finally embraces all the possible imperfections. Regarding this last step, much work remains to be done.

Irregular cases—the ones Stéphane calls "monsters"—remain especially intriguing. Could rules of step-by-step construction be deciphered in these special cases, or is each case different? Investigation is needed to understand the conditions that made them appear. Did growth accelerate too fast? What conditions did the plant experience to create these growth curves? The Araceae family provides a particularly interesting example, as changes occur very quickly, from a single leaf encasing the whole apex to many little flower stems stacked together. In such cases of extremely fast transitions, many species form quasi-symmetric flower stems, as expected. However, some species present perfect Fibonacci stems. How does this happen? Are we missing some hidden slow transitions?

These questions draw attention to another important split that runs through all the work discussed in this book: the tension between "imperfect" reality and

**FIG. 20.1** Two species in the Araceae family pose a mystery. At left, after a quick transition from a spathe, the spadix of a peace lily shows the expected defects of quasi-symmetric phyllotaxis. But at right, the spadix of *Anthurium* displays perfectly regular Fibonacci phyllotaxis.

the "perfect" abstract idealization it inspires in us. For the Pythagoreans, reality was the imperfect realization of pure ideas. But there's another way to view this split: as evidence that our human brains have only limited ability to embrace the full complexity of the world and in fact have a tendency to simplify it. Yes, our observations are solid enough for us to imagine perfection. But our ability to imagine such perfection itself rests on the dynamical processes underlying the development and function of our brains. The world grows step by step, dynamically, with simple interactions leading to increasing complexity. Within this complexity, some general trends and structures can be read. Our limited but efficient brains are quick to grasp these repetitions and form very useful, practical rules from them. Yet the central mystery remains unsolved: How do regularity and structure keep emerging from all this dynamical evolution—leading to both regular phyllotaxis and to our idealistic brains?

Here we pause to once again ponder a fundamental question: *Why* are regular phyllotaxis patterns observed in plants? In this book, we have answered this question in terms of the "proximal" causes for phyllotaxis patterns: they occur because hormone depletion at the tip of the meristem leads to new organs forming in the largest available gap. From this biological starting point and the geometric constraints explained in chapter 14, we arrive at Fibonacci patterns. But what about the "ultimate" cause of particular phyllotaxis patterns? Maybe there is an evolutionary reason that no one has discovered yet—or maybe there is no ultimate cause at all. Perhaps we keep searching for it simply because our brains persistently dream of perfection.

Finally, we return to the title of this book, which after all poses a question. So, *do* plants know math?

The answer is, they don't have to. Fibonacci phyllotaxis arises naturally from simple geometric rules. No golden-ratio goddess waves her magic wand; instead, plant spirals are created through robust biochemical, biomechanical, and dynamical mechanisms that operate at a local level. In building their spirals, plants obey fundamental laws of science—and why shouldn't that gradual unfolding strike our brains as utterly entrancing and thrillingly beautiful?

CHAPTER 21

# A Spiral Dinner (with Recipes)

Now that you have come this far, it's time to cap off your investigation into the complexities of phyllotaxis with a celebratory dinner. Decorate your table with sunflowers and pinecones. All the recipes serve four.

## 1. Signature Drink

### Phylla Colada

To be extra fancy, you might serve this drink in pineapple shells. Then you can count the spirals formed by the pineapple scales while you're sipping. (See the *Try Your Hand* activity following the introduction.) The parastichy numbers for most pineapples are (5, 8) or (8, 13). Instead of ice cubes, this recipe uses frozen pineapple chunks.

Place all ingredients in a blender. Blend until smooth and foamy. If too thick, add a little more coconut milk.

**INGREDIENTS**

3 cups fresh pineapple, cubed and frozen (from 1 large, ripe pineapple)

⅔ cup light coconut milk

⅓ cup rum (or to taste)

**INGREDIENTS**

1 tablespoon extra-virgin olive oil

1 tablespoon butter

2 medium shallots, quartered (or a small onion)

2-pound red cabbage, sliced through its equator, **only one half cored** and cut into 1″ squares, the other half reserved for display

½ cup apple cider (or unfiltered apple juice)

1 teaspoon apple cider vinegar

Salt and pepper to taste

Plain Greek yogurt for serving (optional)

## 2. Appetizers to Choose From

### Sautéed Red Cabbage

The humble red cabbage, sliced crosswise, reveals a beautiful Fibonacci spiral. This recipe uses half a head of red cabbage. Display the other half to your guests.

In a large skillet, heat oil and butter over medium heat. Cook the shallots for a few minutes until tender, stirring occasionally.

Add the cabbage and cook for a few more minutes until the leaves start to soften, stirring occasionally. Then add the apple cider, salt, and pepper. Reduce heat to medium-low and cover.

Cook until the cabbage is tender, about 12 to 15 minutes, stirring now and then. Off the stove, stir in the apple cider vinegar. Serve topped with Greek yogurt, if you wish.

## Artichokes

Buy an extra artichoke if you'd like to try this exercise. Slice off the raw artichoke just above its base. Cut off most of the hairy "choke" and then carefully remove the remaining hairs. You'll see a phyllotaxis pattern appear on the surface of the artichoke bottom.

Fill a large pot halfway with water and bring the water to a boil. While the water is heating up, trim the bottom and top off the artichokes so they sit flat. Snip off the spiny tips of the leaves.

When the water is boiling, squeeze the juice from one lemon half into the water, then toss in the lemon half and the garlic. Boil the artichokes for 20 to 35 minutes, until you can pull off a leaf easily using tongs. Drain the artichokes and place each on a dinner plate.

Melt the butter over low heat. Squeeze the reserved lemon half into the butter and stir. Place the lemon butter into four little bowls, one for each artichoke.

**INGREDIENTS**

4 artichokes (or 5 if you do the phyllotaxis exercise)
1 lemon, cut into halves
2 cloves garlic
1 stick butter

**INGREDIENTS**

2 bunches asparagus

3 tablespoons olive oil

Fleur de sel (or sea salt), to taste

½ lemon

**Roasted Asparagus**

You may never have noticed that asparagus grows in spirals. Before you cook the spears, pick one to examine. Choose a lower leaf scale as your starting point. Find the next leaf scale that grows directly above it (or nearly so). How many scales are between them?

Move an oven rack to the top third of the oven. Preheat the oven to low broil.

Rinse the asparagus, dry it, and spread the spears on a baking sheet lined with foil. A single layer works best—you might need two pans. Drizzle the spears with olive oil. Sprinkle with fleur de sel. Roast for 5 to 15 minutes (depending on the thickness of the asparagus), turning the spears once, until they are flexible.

If you like your asparagus slightly charred, turn the broiler to high for a minute or so at the end. (Watch to make sure the spears don't burn.)

On a serving plate, drizzle with lemon juice and more olive oil.

## 3. Main Dish

### Romanesco Broccoli with Pine Nuts

Romanesco broccoli was what first drew Stéphane into the world of phyllotaxis. Until he took a close look at a Romanesco, he said, "I never realized there was so much regularity in plants. There are all these Fibonacci numbers all around you, but you don't even notice!" So before you cut it up, take a moment to appreciate its spirals, and the spirals within spirals . . .

Pine nuts are the seeds inside pinecones. Nearly all pinecones display spiral phyllotaxis, although some are too irregular to fit Fibonacci patterns.

**INGREDIENTS**

2 heads Romanesco broccoli, cut into small florets
4 tablespoons pine nuts
4 tablespoons extra-virgin olive oil
4 cloves garlic, diced
½ teaspoon red pepper flakes, or to taste
Salt, to taste
2 teaspoons lemon zest, for serving (optional)
Grated Parmesan, for serving (optional)

In a dry frying pan, sauté the pine nuts over low heat until lightly browned, stirring frequently to avoid burning. Set aside.

Set a large frying pan over medium heat. Add the olive oil, garlic, salt, and red pepper flakes, stirring until the garlic is a bit browned. Add the Romanesco broccoli and sauté until just tender, about 5 minutes. Add a little water if needed.

Stir in the toasted pine nuts. If you wish, sprinkle with lemon zest and grated Parmesan before serving.

**INGREDIENTS**

6 sheets phyllo dough, thawed

1 stick butter, melted

3 cups fresh strawberries, sliced

½ cup unseasoned breadcrumbs

½ cup sugar

Powdered sugar, for serving

This makes two rolls (10 pieces).

## 4. Dessert

### Strawberry Strudel in Phyllo

Observe your strawberries before you cut them. The tiny seeds on strawberries display quasi-symmetric phyllotaxis (see chapter 14). Count the spirals, if you dare. "Counting spirals on strawberries is actually extremely difficult," says Chris. "They don't stick to one parastichy pair." Nonetheless, you can observe that the seeds form spirals, even if irregular ones.

The name "phyllo dough" comes from the same Greek root as phyllotaxis: *phyllo* means "leaf." In the case of the dough, the leaf is of course the thin sheet of pastry.

Preheat oven to 400 degrees.

Unfold the phyllo dough and place a clean dish towel over the sheets so they won't dry out.

On another clean dish towel, lay out one sheet of phyllo. Line up the bottom edge of the dough with the bottom edge of the towel. Drizzle the dough all over with a tablespoon of melted butter. Lay a second sheet on top and drizzle with another tablespoon of butter. Then lay a third sheet on top and drizzle with butter once again.

Next, sprinkle three heaping tablespoons of sugar evenly over the dough.

Along the edge of dough closest to you, sprinkle a light band of breadcrumbs about three inches wide. On top of the breadcrumbs, arrange half the sliced strawberries in an even layer. (Make sure you go all the way to the right and left edges.)

Next—this is the only tricky part—use the towel to flip over the bottom edge of the phyllo. Try to keep the strawberries from spilling out. Then roll up the dough the rest of the way, using your hands to make a tube.

Make two rolls. Place the tubes on a baking sheet lined with parchment paper. Slice them into five pieces each before putting them in the oven. Strudel is difficult to slice after it's baked.

Bake the strudel for 15 to 20 minutes, until nicely browned. Sprinkle with powdered sugar before serving.

(Strudel is best the day you make it. If you want to serve it the next day, don't cover or refrigerate it, as it will get soggy. Reheat before serving, if you wish.)

# APPENDIX

## Chapter 2: Chris's Notes on Leonardo and Trees

Leonardo's first-ever recording of plant patterns appears in the top half of his notebook page, which we looked at in detail. But what about the bottom half? These notes I also find interesting, as they offer Leonardo's insights into tree structure and his emphasis on scientific accuracy for artists.

Figure A.1 shows a close-up of his sketch in the margin of the page, with a translation of his observations:

**FIG. A.1** Leonardo's sketches of tree structures.

**Why, very often, timber has veins that are not straight**

When [new higher] branches [grow] . . . off to one side, then the vigor of the lower branch is diverted to nourish the higher one, even though it may be somewhat lopsided. But if the branches grow evenly, the veins of the main stem will be straight [parallel] and equidistant at every degree of the height of the plant.

Wherefore, O Painter who does not know these laws! To escape criticism from those who understand them, you must represent every thing from nature, and not despise such study, as do those who work [only] for money.[1]

Let's assume that by "veins," Leonardo meant the vessels (xylem) *under* the bark that form the vasculature of plants. It seems remarkable to me that in a text clearly geared to painters, he was admonishing them to go beneath the surface and seek to understand the laws of nature, lest they be shamed by other painters who were more knowledgeable. No matter that at this point in history, Leonardo might have been the only one who actually knew. Indeed, what other painter would bother taking the time to remove the bark of a tree to check the orientation of the xylem?[2]

In addition, other Leonardo notebooks indicate he was likely the first to observe that each ring in a tree trunk represents a year in the tree's life, and also that wider rings correspond to wetter years. The science of dating the age of trees by their rings is now known as dendrochronology. Leonardo wrote:

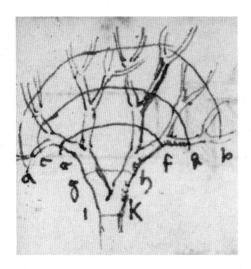

**FIG. A.2** Is there truth in Leonardo's Rule of Trees?

The rings of the branches of trees show how many years they have lived, and their greater or smaller size whether they were damper or drier.[3]

I also find it fascinating to see how Leonardo made observations on vasculature in many forms, connecting disparate elements of nature. Here, for example, he saw the same principle at work in the forms of trees and the flow of water, joining these observations together on the same page of a different notebook:

> All the branches of a tree at every stage of its height when put together are equal in thickness to the trunk [below them].
>
> All the branches of a water [course] at every stage of its course, if they are of equal rapidity, are equal to the body of the main stream.[4]

The first observation is known as Leonardo's Rule of Trees, and (although difficult to prove precisely) it has generally been confirmed by scientists. To restate the rule more simply: if you fold up all of a tree's branches, together they will equal the thickness of the trunk. Much later, scientists would show that this form is fractal-like.[5]

In addition, Leonardo made a direct connection between the vasculatures of trees and human beings, in one case drawing a branching tree right over a human body.[6] Sadly, he missed a major opportunity to advance scientific knowledge by failing to publish his extensive anatomical studies, which he based on dozens of human and animal dissections. This is particularly true of his discovery of atherosclerosis, and also his amazing explanation of the functioning of the aortic valve, confirmed by his glass models of hearts. Both findings would take centuries to be rediscovered.[7]

## Chapter 3: Kepler's Sphere Packing and Planetary Orbits

In his book *The Six-Cornered Snowflake*, Kepler compared the shape of pomegranate seeds to the rhomboid packing of honeycombs made by bees. He observed that in a pomegranate, the seeds are spherical at first but then become rhomboidal under pressure from their neighbors as they grow. This led him to wonder which type of sphere packing would be optimal for fitting the most balls in a given space. Kepler conjectured that the usual grocery store packing of apples or oranges, as in his figure B, would be optimal. A full solution to the problem would not be proven until 1995.

At age 25, Kepler had a geometric-spiritual vision of the solar system, as described in his book *The Cosmographic Mystery*. He saw each planet's orbit as lying on a concentric sphere, or "orb." Each orb's dimensions fit neatly between a succession of hollow Platonic solids, with the sun at the center. The orb of Mercury was an octahedron. The orb of Venus was an icosahedron, while that of Earth was a dodecahedron—both related to the golden ratio. Despite being contradicted by Kepler's own

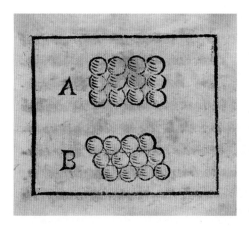 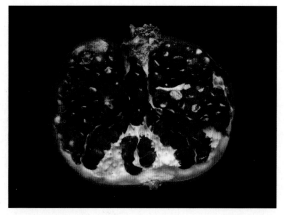

**FIG. A.3** At left, Kepler's illustration of sphere packing. At right, the rhomboid packing of pomegranate seeds.

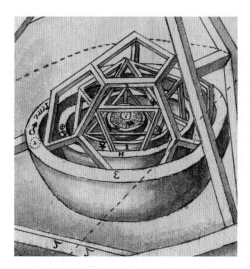 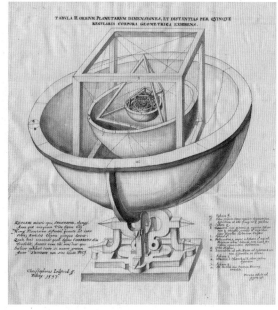

**FIG. A.4** To Kepler, the orbits of the six known planets could be represented as nesting polyhedrons, illustrated in his 1596 book *Mysterium cosmographicum* (*The Cosmographic Mystery*).

First Law, which states that planetary orbits are ellipses, this model gives fairly accurate proportions between each planet's distance from the sun. Kepler's nesting polyhedrons in figure A.4 are drawn in the style of Leonardo.

## Chapter 4: More on Bonnet's Exceptions

Bonnet did not shy away from noting several exceptions and amendments to his five orders. Not only did this reveal his honesty as an observer, but these details made his work more universal and ahead of its time.

- **Quincunx exceptions.** "The fourth order has its variations," Bonnet wrote.[8] Instead of growing in the usual groups of five, he noted, sometimes the spiral pattern repeats in groups of three, seven, or eight leaves.[9] All told, he looked at 64 specimens from 12 different plant species. He found four cases, mainly in peach trees, whose leaf pattern repeats in groups of three. He found a single case of leaves grouped by seven. And he found five cases, all in apricot trees, of leaves grouped by eight. Note that the most frequent numbers to appear are the Fibonacci numbers 3 and 8. The plot thickens when you realize that Bonnet's fifth order can also be viewed as a quincunx exception, with the pattern repeating every 21 or 55 leaves.[10]

- **Twisting of the quincunx.** Bonnet observed that the vertical rows joining every five leaves in the quincunx were not so vertical after all.[11] He wrote:

  > One day as I was observing a shoot from an Apricot tree, I noticed that the first leaf of the second quincunx deviated a bit to the right of the line where the first leaf of the first quincunx appeared. I noticed that this deviation continued in the same direction, & following the same proportion, in all the corresponding leaves; & that it formed a spiral winding around the stem. At first, I suspected that this was one of those special cases from which one can draw no conclusion. But having examined many other shoots of the same species, & shoots from other species, I saw in each the same peculiarity, the same deviation. Sometimes the spiral wound up from right to left, sometimes from left to right.[12]

  Bonnet found this observation "extremely pleasing," as he believed it supported his conviction that leaves avoid overlapping in order to better pump dew and breathe. This nonalignment of leaves would become a crucial hypothesis of the Bravais brothers (see chapter 7).

- **Plants displaying two orders** (see fig A.5). Bonnet noted that two orders can appear in the same specimen, but "modest attention is sufficient to determine the correct one."[13] He added that some plants truly embrace two orders. Myrtle and pomegranate trees may display both the second order (decussate) and the third (whorled). Hemp may have lower leaves arranged according to the second order (decussate) and upper leaves according to the fourth (quincunx).

- **Corn complications.** In his observations of 724 corncobs, Bonnet looked at vertical (or slightly spiraling) rows of kernels.[14] Many were more irregular at the base than at the top, where rows could be distinguished. Among the regular cobs, he counted 3 with 8 rows; 252 with 12 rows; 44 with 14 rows; 78 with 16 rows; and 16 with 18 rows. Apart from commenting on the changing shape of the kernels between bottom and top, Bonnet did not seem to try to fit these structures into his five orders. (This mystery, which would not be solved until the twenty-first century, is the first mention of patterns now called "quasi-symmetric." See chapter 14.)

- **Environmental factors.** See figure A.6.

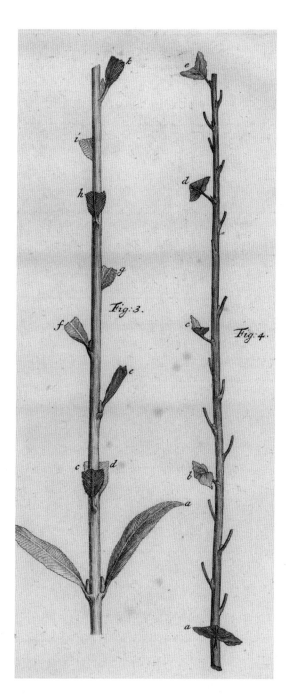

**FIG. A.5** Here, Bonnet shows plants displaying two different orders. On the left, the plant has at its base two decussate pairs—a, b and c, d. Above them is what Bonnet calls a quincunx, formed by e, f, g, h, i, and k.

At right, the apricot branch shows a transition from a decussate pair at bottom to a more straightforward quincunx structure. But the leaves at the start of each quincunx—a, b, c, d, and e—are not entirely aligned vertically, instead spiraling around the stem. Bonnet found this "extremely pleasing," as this non-alignment of leaves supported his conviction that leaves avoid overlap in order to better pump dew from the ground. He observed, in dozens of plants, this spiraling going in either direction—clockwise or counterclockwise.

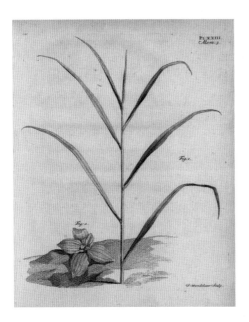

**FIG. A.6** Bonnet observed that if a plant grows next to a wall, its normal leaf arrangement (at left) may be disrupted (as at right), except at the points where the leaves attach.

## Chapter 5: More on Fraction Trees

The elegant "trees" that Karl Schimper and Alexander Braun created in the early 1800s exist in many variations. Perhaps the best known now are the Stern-Brocot tree and the Farey sequence. All are based on rational numbers, adding the new term as the "mediant" (not the mean) between two consecutive rational fractions. But over the centuries, mathematicians have chosen different ways of ordering the information.

- Computation of the mediant dates back to the French mathematician **Nicolas Chuquet** in 1484.
- Another French mathematician, **Charles Haros**, used the mediant in 1802 to create a list of all rational numbers with denominators smaller than 100, looking for ways to ease the country's transition to the metric system. He started from $1, \frac{1}{2}, \frac{1}{3}, \ldots$, down to $\frac{1}{100}$. To make computations faster, he added the symmetric numbers starting from $\frac{1}{2}$: $1-1, 1-\frac{1}{2}, 1-\frac{1}{3}, 1-\frac{1}{4}, \ldots$ Then he took the mediant between terms, stopping when he reached a denominator larger than 100. (He did not draw trees.) He demonstrated that two consecutive terms, such as $\frac{p}{m}$ and $\frac{q}{n}$, are "relatively prime," meaning they have the property that $np - mq = 1$.

◎ In 1816, British geologist **John Farey Sr.** looked at a similar sequence, limited up to a larger denominator (1,024). He conjectured that the new elements are mediants of the surrounding pair, and that two consecutive fractions are "relatively prime." This was demonstrated soon after by **Augustin-Louis Cauchy**, who coined the term "Farey sequence"—though neither he nor Farey knew what Haros had already done.[15] Represented as a tree, the fractions are organized from the top down, starting with two integers and then inserting all the medians in succession, stopping, as Haros did, when the numerator reached a certain limit.

◎ In 1830, **Schimper** made his table of mediant values (see fig. 5.6), with the new mediant value being placed between the preceding pair, starting with $\left[\frac{1}{2}, \frac{1}{3}\right]$ and $\left[\frac{1}{3}, \frac{1}{4}\right]$. He organized the table from the top down, each horizontal line displaying fractions of fixed *numerator* (each line resulting in an increasing denominator). The fractions already introduced appeared on subsequent lines in their nonreduced form (for instance, $\frac{1}{2}$ reappeared as $\frac{2}{4}, \frac{3}{6}, ...$). This marked the first time the Haros-Farey sequence was represented as a table or tree.

◎ In 1831, **Braun** made a plot of the values, placed precisely along the segment $\left[\frac{0}{1}, \frac{1}{1}\right]$ (see fig. 6.6). He organized the plot from the top down according to fixed *increasing denominators*, as is now standard. Like Schimper, he also used nonreduced fractions, but he placed them all in their exact positions along the interval. Braun's tree was very close to the Haros-Farey sequences, as both allowed the sequence of fractions to stop at a given denominator or required precision (like $\frac{1}{10}$ or $\frac{1}{100}$, as in the original Haros sequence).

◎ In 1858, the German number theorist **Moritz Abraham Stern** rediscovered the same infinite complete binary tree. Independently in 1861, the French clockmaker **Achille Brocot** created the same tree for a practical purpose: to help him design systems of gears. They both organized

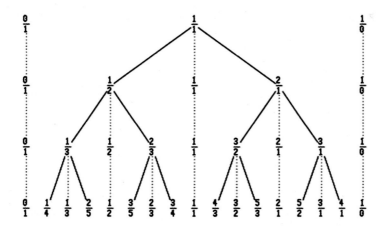

**FIG. A.7** The Stern-Brocot tree, with Stern-Brocot sequences of order *i* for *i* = 1, 2, 3, 4.

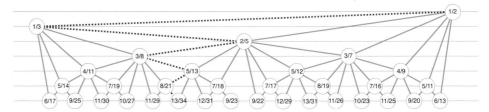

**FIG. A.8** A variation on Schimper's table, this time with the first terms $= \begin{bmatrix} \frac{1}{3}, \frac{1}{2} \end{bmatrix}$ increasing in value and ordered by mediant operations, as in Stern-Brocot. Note how the Fibonacci fractions, joined by red dashes, form a familiar zigzag. They also display the highest numerator and denominator, for each line.

their trees starting from $\frac{0}{1}$ to the symmetric $\frac{1}{0}$ at the top, adding to the sequence according to the number *of mediant operations*.[16]

From this detailed analysis of trees, we see how botany introduced some very original mathematical ideas on fractions. Schimper's tree was the first to show in table form how to produce a sequence of fractions, starting from only a pair of fractions and repeatedly finding the mediant. Schimper was also first to order the fractions logically according to the numerator (with the denominators necessarily decreasing as the fractions increased).

Schimper's botanical example provided an intuitive way to understand the origin of the mediant computation (a weighted mean value of the divergence angle over full periods). And finally, Schimper introduced the use of nonreduced fractions, drawing connections to fractions higher up in the table. Stéphane notes: "Unfortunately, Schimper's profound insight—showing the formation of not only all the reduced fractions but also all the nonreduced ones—has yet to catch on among mathematicians."

## Chapter 6: More on Braun's Geometry and Vectors

Alexander Braun found surprisingly complex relationships in the spirals on a pinecone. Here, we present some of his mathematical observations in more detail.

In figure A.9, Braun shows how to construct both flatter and more vertical spirals on an idealized pinecone as a technique for finding the divergence angle. Here, each scale is shown as a four-sided tile. Starting at *c* (on the left, next to *a*), first we see the two obvious spirals: the left-handed spiral that passes through *e, d, d,* and the right-handed one passing through *e, e, e.*

Next, we can define two more spirals, one flatter and one more vertical. The flatter one—*c, e, b*—is indicated by a thin solid line. The more vertical spiral—*e, f, f*—is indicated by a delicate dotted line. These spirals are not yet perfectly horizontal and vertical, so Braun must repeat the process at least once more.

Next, we draw the even flatter diagonal that passes through *e* and *a*, marked by the bold dotted line: this is the generative spiral we were looking for. Similarly, we find the most vertical diagonal, which passes through *e, g, g,* as indicated by the dashed line. Once we have the generative spiral and the most vertical diagonal, we can find the fractional value of the divergence angle.

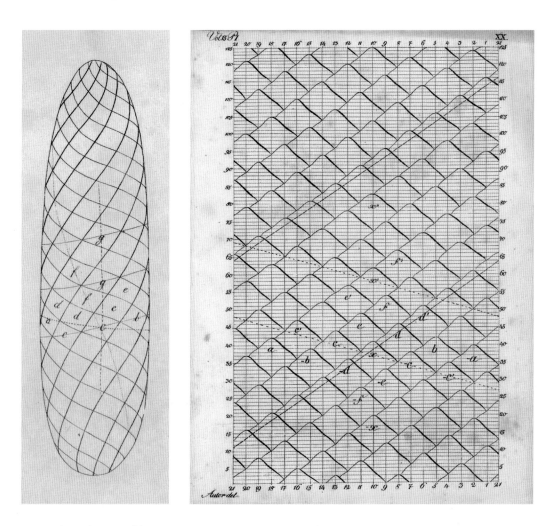

**FIG. A.9** A mathematical lesson in pinecone spirals, from Braun.

**FIG. A.10** Braun's "unrolled" pinecone, a 2D view of a 3D cone.

In figure A.10, Braun shows another view of the same system, this time as if the pinecone had been "unrolled" on a plane, making one full rotation. There are three left-handed spirals, following the angle of the *dotted lines*. The left-handed spiral that passes through -*c'*, -*c*, *x*, *c*, *c'* reappears three spirals above, passing through *x'*, indicating that we need three such parallel spirals to tile the whole cylinder.

In the other direction, there are five right-handed spirals, following the angle of the *dashed lines*. The spiral that passes through -*d*, *x*, *d*, *d'* reappears five parallel spirals higher (above *x''*), indicating that we need five such right-handed spirals to tile the whole cylinder. Note that the vertical orthostichy appears at -*x*, *x*, *x'*, *x''*. Also note that the generative spiral passes through -*a*, *x*, *a*, *b*, *c*, (unlabeled scale), *d*, *c'*.

Taking another intellectual leap, Braun saw that progressing toward flatter spirals corresponds to the combination of vectors (see fig. A.10). The two original vectors are *x*, -*d* and *x*, *c*. Adding them together, you get the new vector *x*, -*b*. The next combination adds this one to the previous vector *x*, *c*, leading to the new vector *x*, *a*.

Repeating this process, you add to the vector *x*, -*b*, and this leads to -*a*. Now we have returned to the same spiral—but reversed. This is a characteristic of the generative spiral. Another way to see the relationship is to subtract the number of spirals: since there are 5 and 3 original spirals, this gives us 2 for spiral *x*, -*b*. Then 3–2 gives us 1 for the spiral *x*, *a*, which means it is the generative spiral. The next step, 2–1, again gives us 1—meaning it is the same generative spiral that we found using the vector combination.

In figure A.11, Braun's vectors show the directions of the spirals discussed in the idealized pinecone from figure A.9. Here, Braun's drawing has been relabeled according to modern conventions. With the first scale at 0, the two obvious spirals connect 0 to 2 and 0 to 3. This means the pinecone has two right-handed spirals and three left-handed ones.

Next, we find both the flattest and the most vertical diagonals. We start by combining the vectors 0 → 3 minus 0 → 2. The resulting vector from 0 → to 3−2 = 1. When the result is 1, this means we have reached the first scale and have found the generative spiral.

Now we repeat the process on the other side, which leads us to the same point 1 on the left. Therefore the distance between 1 at right and 1 at left is the whole circumference of the cylinder unrolled. To obtain the other diagonal, we add 0 → 2 to 0 → 3 (indicated by the dashed arrow heading up from 3). This gives the vector 0 → 2+3 = 5, which is not yet vertical. We then repeat the same operation using the sum of the two last vectors, similar to the Fibonacci addition process. Adding 0 → 3 to 0 → 5, we get the vector from 0 → 3+5 = 8, which is now drawn vertically. These vector combinations lie at the heart of the Bravais brothers' work.

Braun's drawing in figure A.12 resembles modern views of the law of combination of vectors. The detail shown here corresponds to Schimper's table between 1/3 and 1/2. We have connected the points on Braun's plot using red dotted lines to show the golden angle divergences, which form a distinctive zigzag.

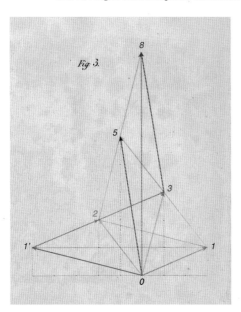

**FIG. A.11** Braun's demonstration of vector combinations.

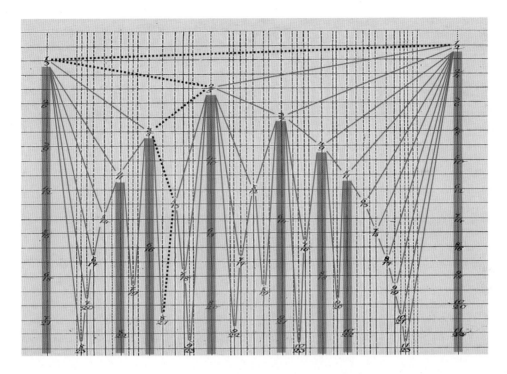

**FIG. A.12** Connecting Braun's dots. The detail shown here corresponds to Schimper's table between 1/3 and 1/2. We have connected the points on Braun's plot using red dotted lines to show the golden angle divergences, which form a familiar zigzag.

## Chapter 12: More on Levitov's Energy Diagrams

In figure A.13, levels of Levitov's energy function appear in boldface, with arcs corresponding to regular spiral lattices on a plant. In the diagram above, the little circles inside the triangles are energy minima, corresponding to compact hexagonal packing. The lighter half-circles start from the "mountain summits" on the x-axis (with endpoints 1, 0.8, 0.75, . . . , 0.5), representing energy maxima. These arcs correspond to rectangular lattices. The arcs descend to the "mountain passes" between the triangles, at points that correspond to square lattices. These semicircles delineate triangular regions that map onto one another under the symmetries of the repulsion function.

In figure A.14, the crooked curves represent (local) minima of energy for each fixed $y$, the magnetic compression value. This is just like the compression value in the Bravais lattices and matches the relative size of Van Iterson's primordia. The crooked curves rise up to a "mountain pass" where they cross the arcs of circles (and where the lattice is square). Then they head sharply down to the small triangular dips (where the lattice is hexagonal). These crooked curves are identical to the branches in the Van Iterson diagram and have the same symmetries. But now they are disconnected.

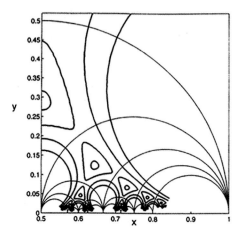

**FIG. A.13** Semicircles in Levitov's energy diagram.

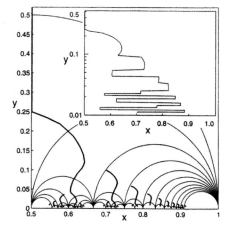

**FIG. A.14** Crooked curves in Levitov's energy diagram.

The inset graph shows the global minimum of energy as $y$ decreases, jumping from one part of a branch to another. It reveals that the Van Iterson branch is a minimum of repulsion energy when we continuously deform the pattern by compressing it—but not necessarily the *best* repulsion minimum when we look in absolute terms to find the least repulsive lattice.

## Chapter 13: Stéphane's Calculations, Explained

Over the course of just one year, Stéphane and his boss Yves Couder published three papers on Fibonacci phyllotaxis. The following details shed light on what Stéphane was thinking at the time, as he used physics and computers to attack the problem.

### MORE ON FIGURE 13.7

In the version of Stéphane's model in figure 13.7, the primordia form periodically at the apex and then move away at constant velocity as the shoot grows. (The model follows Hofmeister's rule that

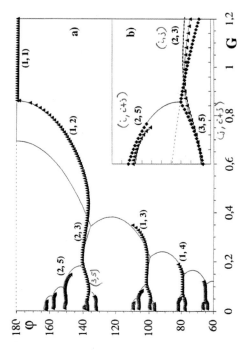

**FIG. A.15** One of Stéphane's early computer models.

each leaf arises in the largest gap between previous leaves.) The x-axis is the dynamical parameter $G$: the ratio of the distance traveled by the primordium during time period $T$ to the radius of the apex. The y-axis is the divergence angle $\varphi$.

For each value of $G$, Stéphane ran a computer simulation, plotting a point $(G, \varphi)$ if the simulation converged to a lattice with the divergence angle $\varphi$. Just as in Van Iterson's diagram, the different branches correspond to lattices with parastichy numbers $(m, n)$. In the inset, they are also indicated generally as

$$(i, j) \text{ or } (i, i+j).$$

To obtain the diagrams, Stéphane would first run a simulation that produced slightly irregular states. Not knowing how to treat them (yet), he would average the divergences obtained and then restart the simulation using a regular system with these average divergences. Ultimately, this would converge toward a perfect regular system whose parameters he would record and plot. Then he would slightly change the growth parameter and restart the process.

To draw the separate branches, Stéphane had to start with a theoretical regular arrangement close to the expected result and then repeat the same convergence process, increasing and decreasing $G$ to draw the whole branch. At the uppermost point of a branch, the pattern would break and jump (irregularly) to another possible branch.

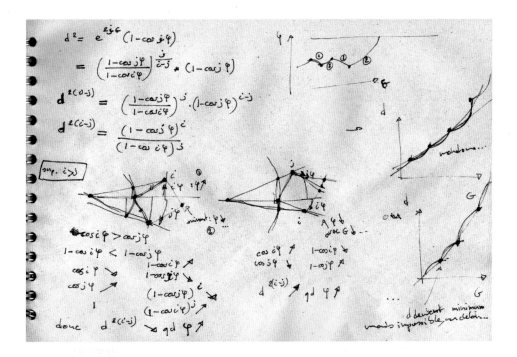

**FIG. A.16** Stéphane's thinking in ink.

## MORE ON FIGURE 13.8

In figure 13.8, Stéphane explained notes he made when he was trying to solve the puzzle of Fibonacci phyllotaxis using physics. At top right on the page, he sketched the basic oscillation of the divergence angle along a Fibonacci branch (with the x-axis being the dynamical parameter $G$, and the y-axis the divergence angle $\varphi$).

Drawings in the middle of the page show the movement of the primordia in the "good direction" that will generate Fibonacci spirals. In the two "diamonds," the lines on the left side cross at the center of the apex, with the central zone indicated by an arc of a circle. The new point 0 lies on this arc, in contact with the two points $i$ and $j$. Here, it is supposed that $i>j$, as noted in the little box. This means that point $i$ is farther away from the new point, so the rhombuses with left corner angles $i$-0-$j$ are tilted. Therefore, the projection on the arc of the circle shows a small length for $i$ and a large one for $j$.

The two diamonds also show two cases for $i$: at the top and on the bottom. The arrows indicate the direction of variation in the positions and corresponding divergence, following the diagram and the computation. Stéphane checked that these matched the oscillation of the Van Iterson branch, sketched at top right.

## MORE ON FIGURE 13.9

In figure 13.9, Stéphane was trying to figure out how disk 5 could find the *best possible* minimum in order to fulfill the requirements of dynamical stability. Initially, disk 5 sat almost directly above disk 0 and in contact with disk 3, following the regular arrangement supposed. But because both supporting disks were then on the same side, this was not a stable position. (Try standing with both feet on the same side of your waist!) A stable position requires being seated between two disks on opposite sides. Here, disk 5 will roll into the lower position 5*, indicated by a dotted circle. And so we see that this (2, 5) construction—which was assumed geometrically in the Van Iterson diagram—cannot be grown dynamically using this parameter, because the first disk placed would fall out of line.

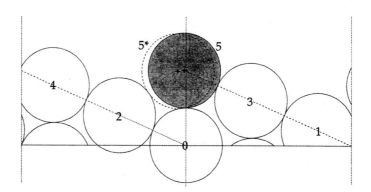

**FIG. A.17** Searching for stability. Stéphane's threshold is not the same as either Levitov's or Adler's.

Trying again using slightly smaller disks, Stéphane found that the two links 5-0 and 0-2 would run in opposite directions. This time, the original position of disk 5 becomes a minimum. The transition occurs when disk 5 sits vertically above disk 0, which is Adler's threshold on the Van Iterson arc. But this case still would not work in the dynamical model.

To fulfill the requirements of the dynamical model, the new position needs to be not only a minimum but the *best possible* minimum. The original position of disk 5 must still be compared to the disrupting disk at 5*, sitting between 0 and 2. Along the Van Iterson arc, position 5 (always between 0 and 3) would indeed be the best minimum, lower than 5*, but still after the threshold that Stéphane found to be the best for dynamical stability (as shown in fig. 13.10). Ultimately, the diagram in figure 13.9 helped Stéphane see that while his threshold of dynamical stability always falls *after* Irving Adler's vertical threshold, it can fall either *before or after* Leonid Levitov's square threshold.

## MORE ON FIGURE 13.12

In figure A.18, Stéphane's printout shows results based on the Snow and Snow hypothesis. This model has the advantage of allowing formation of several primordia at the same time, which can lead to more possibilities. Here, the parastichy numbers are bijugate, indicating a plant with leaflets opposite each other on a stem. The parastichy numbers for bijugate plants are represented by doubling the Fibonacci ones: in this case $2 \times (2, 3)$, with some noticeable irregularity.

At top left on the page, the mean divergence angle $i$ is indicated, as well as its decomposition as a continuous fraction (meaningless in this irregular case). At right are the simulation parameters, followed by the last successive divergence angles. Beneath the spirals, a horizontal plot shows the successive divergence angles, with the "vampire teeth" cut back, because of the folding of π (−180 = +180) as indicated. Stéphane has also sketched a figure related to the order of primordia appearance, looking for the right succession of divergence angles.

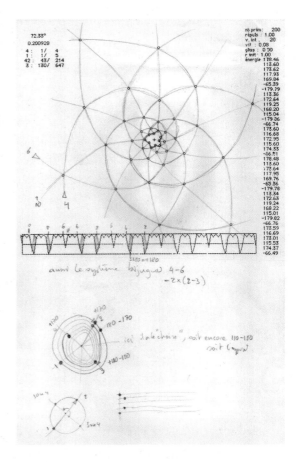

**FIG. A.18** Here, Stéphane's data show the results of a computer simulation he ran that looked at the Snow and Snow hypothesis, in which new leaves emerge not only *where* there is the biggest gap but also *when*.

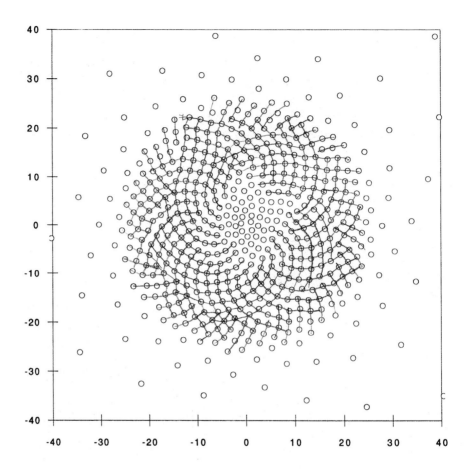

**FIG. A.19** This was Stéphane's first computer simulation of a sunflower, annotated by hand. The parastichy numbers are (34, 55) toward the outside, changing to (21, 34). Again, one finds perfect Fibonacci numbers, but the pattern itself shows some irregularities.

## Chapter 14: More on Zigzag Fronts

Here, we zoom in on the Mandelbrot set spirals that first drew Chris to phyllotaxis, and we also consider different interpretations of the disk-stacking data.

### SPIRALS INSIDE THE MANDELBROT SET

In this version of the Mandelbrot set, the spirals are obtained by repeatedly applying the function $f(z) = z^2 + c$, where $z$ is a complex variable, and $c = a + ib$ is a complex number represented by a point $(a, b)$ in the main body of the Mandelbrot set. The spiral converges to a fixed point $f$. In the figure, three starting points $c$ are circled in blue, each leading to a spiral structure.

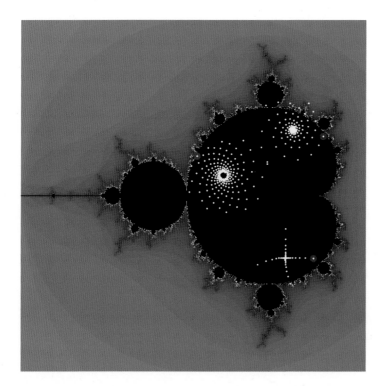

**FIG. A.20** A Mandelbrot set with three related spiral patterns. In a graduate class taught by Chris, Scott Hotton partitioned the Mandelbrot set into regions according to the parastichy numbers of the spiral at each point.

## WHAT DO DIGITAL SIGNATURES TELL US?

The three graphs in figure A.21 show digital signatures of Fibonacci and quasi-symmetric transitions, using data from disk stacking simulations. At top left, as disks are stacked, successive fronts are formed and their parastichy number pairs plotted. The Fibonacci transitions display a characteristic series of interweaving plateaus at the heights of successive Fibonacci numbers. The green and red graphs head upward when triangle transitions occur. At bottom left, in the quasi-symmetric case, the two graphs closely follow one another as the triangle transitions constantly change sides, as seen in figure A.22.

In the right-hand image of figure A.21, the orange and blue graphs represent the quotient of the largest over the smallest parastichy numbers for each front. The Fibonacci case appears in orange, with plateaus close to the golden mean. The quasi-symmetric case appears in blue, with the quotient tending to 1.

Although Stéphane found the full fronts very useful, he felt that to visualize how they changed during disk stacking, he needed to reduce each front to a point. To do this, he made a projection using the sum of only the up vectors of a front, without the down vectors. For Fibonacci phyllotaxis, the resultant vector (sum of the up vectors) forms a butterfly as the disks stack up. It goes up diagonally with triangle transitions on one side, goes down slowly as the front flattens, and up again diagonally

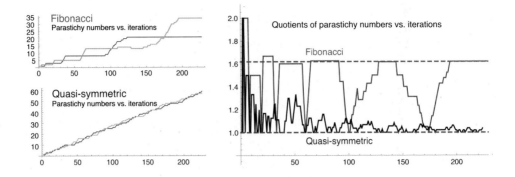

**FIG. A.21** Graphs generated using data from disk stacking simulations.

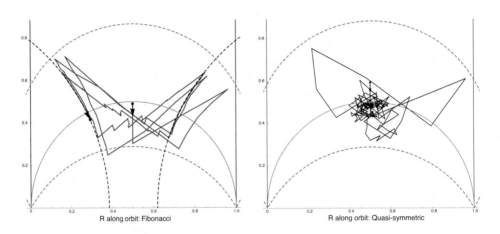

**FIG. A.22** When Stéphane made projections of Chris's disk stacking simulations, the graph for Fibonacci phyllotaxis formed a butterfly (at left), while the graph for quasi-symmetric phyllotaxis resembled more of zooming fly (at right).

in the opposite direction with triangle transitions on the other side. The dotted upper and lower arcs are the hexagonal states limits, the central orange arc represents square states, and the side arcs are for triangles of golden ratio proportions.

For quasi-symmetric states, by contrast, the resultant vector moves more randomly toward the center, with a quasi-identical number of spirals tending toward the orange line of the square state.

## Chapter 15: A Closer Look at Fractals

For enthusiasts of fractals and hyperbolic geometry, the following discussion gets into more of the nitty-gritty.

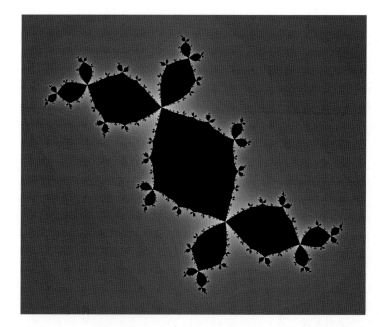

**FIG. A.23** The Douady rabbit.

## MORE ON FIGURE 15.5

The Douady rabbit in figure A.23 represents a set of complex points $z_0$ whose iterations under the recursive rule $z_{n+1} = z_n^2 + c$ ($c$ is a complex number and $n = 0, 1, 2, \ldots$) do not go to infinity as $n$ goes to infinity. These are called Julia sets. In this case, $c$ has a specific definition.

The value that Adrien Douady liked to use for $c$ was the solution of $(c^2 + c)^2 + c = 0$, for which the rule has a cycle of three points. (Here, the value of $c$ approximately equals $-0.12 + 0.75i$.)

## MORE ON THE VAN ITERSON FRACTAL DEBATE

Although the Van Iterson tree iterates to infinity, it preserves its smooth construction history. These smooth curves cannot be considered a fractal object because of their smoothness—even if they show a nice self-similarity.

Mandelbrot's way of getting out of this conundrum was to define a fractal not by its self-similarity, but with a more abstract definition. A line segment can be constructed from smaller segments by self-similarity, but of course it can also be drawn directly. By contrast, "good fractals" cannot be drawn except by infinitely iterating a simple step. The result is a highly complicated structure, even if it's continuous like the Koch snowflake. This led Mandelbrot to a more demanding definition—stating that in order to be a fractal, the Hausdorff dimension (roughly the dimension of the density of points) should be different from the topological dimension (the dimension of the links of points).

If we start with the Koch curve, the similarity of two pieces of the curve can be checked by rescaling, rotating, and translating one so that it can perfectly fit the other. But the similarities in the

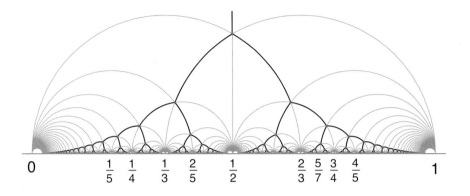

**FIG. A.24** A fuller view of the Van Iterson diagram.

Van Iterson tree belong to another geometry, that of hyperbolic geometry. In hyperbolic geometry, the shortest path between two points (a.k.a. a geodesic) is either an arc of a circle perpendicular to the x-axis, or a vertical line.[17] In that sense the Van Iterson diagram is made purely of these geodesics. The similarities, which here are hyperbolic isometries, are of the form $z \to (az+b)/(cz+d)$ where $a, b, c, d$ are integers satisfying $ad - bc = 1$, and $z$ is a complex number. As a result, every arc in the diagram has the same (hyperbolic) length.

The self-similarity of the Van Iterson diagram is an expression of the similarity of the meshes of the lattices that the diagram represents. For instance, points of the diagram corresponding to lattices with a square mesh are mapped onto one another under these similarities. This is the basic idea of renormalization. (Note that another type of renormalization ties the local similarities of trajectories of the rule $z_{n+1} = z_n^2 + c$ to the structure and self-similarities of the Mandelbrot set.)

The Italian mathematician Eugenio Beltrami came up with the "upper half-plane" model of hyperbolic geometry in which the Van Iterson diagram resides. Some say that whales live in an upside-down version of this model, with the x-axis representing the surface of the ocean: because sound travels faster in denser, deeper water, it tends to follow hyperbolic geodesics. Beltrami also introduced a disk model in which geodesics are arcs of circles perpendicular to the bounding circle. (Stigler's law of eponymy strikes again: these two models now bear the name of Poincaré.) Whatever its name, this disk model more than once inspired the artist M. C. Escher.

**FIG. A.25** A Poincaré (Beltrami) hyperbolic disk, which is self-similar.

# NOTES

## Notes to Introduction

1. Oddly enough, there is no consensus on how to pronounce the first syllable in phyllotaxis. American dictionaries say "fill," this book's authors say "fyle," and the British generally say "feel." Everyone agrees, however, that the stress falls on the third syllable, phyl-lo-TAX-is.
2. See Brian Butterworth, C. R. Gallistel, and Giorgio Vallotigara, "Introduction: The Origins of Numerical Abilities," *Philosophical Transactions of the Royal Society B* 373, no. 1740 (2018): 1–4.
3. Try your own experiment in leaf mobility: if you turn a houseplant in a sunny window by 180°, at first it will appear strangely backward. But after just a week or so, the leaves will have turned toward the new direction of the light.
4. Technically speaking, a spiral in three-dimensional space is called a helix. The curves winding up a pineapple or a closed pinecone, for example, are helices. Following phyllotaxis conventions, however, we will use the term "spiral" when counting and analyzing these curves.

## Notes to Chapter 1

1. A Fibonacci poem follows the Fibonacci sequence in the number of syllables per line. We will use these little poems to introduce our chapters, each having a last line of 13 syllables.
2. Note that the Egyptian white lotus, *Nymphaea lotus*, is not actually a lotus but instead a type of water lily.
3. Theophrastus, *Enquiry into Plants*, trans. Sir Arthur Hort (Cambridge, MA: Harvard University Press, 1990), vol. 1, 75.
4. Theophrastus, *Enquiry into Plants*, 11.
5. Han Ying, as quoted in Philip Ball, *Nature's Patterns* (Oxford: Oxford University Press, 2009), 2.
6. Ball, *Nature's Patterns*.
7. Pliny, *The Natural History of Pliny*, vol. 5, trans. John Bostock and H. T. Riley (London: Henry G. Bohn, 1856), 227.
8. This is actually a pair of real leaves (and axillary bud branches), and so whorled by two like myrtle, but each with two or three neighboring false leaves.

9. Fibonacci, introduction to *Liber abaci*, as quoted in Keith Devlin, *The Man of Numbers: Fibonacci's Arithmetic Revolution* (New York: Walker, 2011), 39.
10. A better translation of *Liber abaci* would be "Book of Calculations."
11. Devlin, *Man of Numbers*, 3.
12. Fibonacci, as translated by Laurence Sigler in *Fibonacci's Liber Abaci* (New York: Springer Science+Business Media, 2002), 15.
13. See Roshdi Rashed, "Fibonacci et les mathématiques arabes," *Micrologus: Nature, Sciences and Medieval Societies* 2 (1994): 145–60.
14. During the "Dark Ages" of Europe, Arab mathematicians wrote many original works based on the Greek mathematicians to whom they still had access, as well as the Indian mathematicians with whom they were in constant contact.
15. See J. J. O'Connor and E. F. Robertson, "Leonardo Pisano Fibonacci," MacTutor, University of St. Andrews, https://mathshistory.st-andrews.ac.uk/Biographies/Fibonacci/, and Frances Carney Gies, "Fibonacci: Italian Mathematician," Britannica, http://www.britannica.com/biography/Leonardo-Pisano.
16. Translations of these passages from Fibonacci's *Liber abaci* have been edited by the authors.
17. The favor was returned to Lucas: mathematicians gave the name "Lucas sequence" to the Fibonacci-like sequence that begins 1, 3, 4, 7, 11, . . .
18. Some experts argue that knowledge of the sequence dates all the way back to Pingala, who wrote the earliest known treatise on Sanskrit prosody in the third or second century BCE. His language is challenging to interpret, however, even for scholars. As translated by Jayant Shah, a math professor at Northeastern University, Pingala described the sequence this way: "The count [of all possible forms of a *mātrā* meter] is obtained by adding [the counts of] permutations of the two previous [*mātrā* meters]. This is the way to the count [of total permutations] of succeeding *mātrās* [*mātrāmeters*]." According to Shah's interpretation: "Thus, we get the formula $S_n = S_{(n-1)} + S_{(n-2)}$, which generates what is now known as the Fibonacci sequence." Jayant Shah, "A History of Piṅgala's Combinatorics," 38–39, http://www.northeastern.edu/shah/papers/Pingala.pdf.
19. Parmanand Singh, "The So-Called Fibonacci Numbers in Ancient and Medieval India," *Historia mathematica* 12, no. 3 (1985): 234. Hemachandra's treatise on Sanskrit poetic meters is titled *Chandonuśāsana*.
20. Singh, "So-Called Fibonacci Numbers." Translation edited for clarity.
21. See Singh, "So-Called Fibonacci Numbers."

## Notes to Chapter 2

1. Da Vinci was not his family name—those were rare at the time—but simply indicates that Leonardo came from Vinci, a small town near Florence.
2. Leonardo da Vinci, *The Notebooks of Leonardo da Vinci*, ed. Jean Paul Richter (New York: Dover, 1970), vol. 1, 207 (section 402).
3. Leonardo's student Francesco Melzi, who inherited his teacher's notebooks and many of his belongings, compiled Leonardo's notes on anatomy, landscape, and art around 1540. A century later, these were published in France; the English version is known as *A Treatise on Painting*.

4. Leonardo da Vinci, *The Literary Works of Leonardo da Vinci*, ed. Jean Paul Richter, bilingual ed. (London: Sampson Low, Marston, Searle and Rivington, 1883), vol. 1, 211. This passage appears in the "Botany for Painters" chapter, section 412.
5. Leonardo's observation is arguably the first inference that side branches sprout from just above the leaves, from what are called "axillary" buds.
6. Leonardo da Vinci, *Literary Works*, vol. 1, 208 (section 404). Italics added. Leonardo even mentions that leaves turn in the opposite direction on the axillary stem from the main stem.
7. In modern terms, this would (presumably) be (2, 3) phyllotaxis, with two spirals in one direction and three in the other.
8. If only Leonardo had published his notebooks!

## Notes to Chapter 3

1. Johannes Kepler, *Mysterium Cosmographicum: The Secret of the Universe*, trans. of 1596 edition by A. M. Duncan (New York: Abaris Books, 1981), 133. In the original Latin, Kepler calls the golden section "de sectione proportionali." The often-quoted second half of Kepler's observation, comparing the proportion to a "jewel," appears several pages later in Kepler's book. (In the Duncan translation, it appears on p. 143).
2. Albert van der Schoot, "Kepler's Search for Form and Proportion," *Renaissance Studies* 15, no. 1 (2001): 60n6.
3. The original Latin title of Kepler's booklet is *Strena seu de nive sexangula*. A *strena* was a New Year's gift.
4. Johannes Kepler, *The Six-Cornered Snowflake*, trans. Colin Hardie (Oxford: Oxford University Press, 2014), 21.
5. The convergence of quotients of Fibonacci numbers to $\phi$ is slow—in fact it is the slowest possible among all rational approximations of irrational numbers. In this sense, $\phi$ is the most irrational number.
6. For another view, see Leonard Curchin and Roger Herz-Fischler, "De quand date le premier rapprochement entre la suite de Fibonacci et la division en extrême et moyenne raison?" (When was the first link between the Fibonacci sequence and the division in extreme and mean ratio?), *Centaurus* 28, no. 2 (1985): 129–38. The authors reproduce a handwritten margin note they found in a copy of Luca Pacioli's 1509 translation of Euclid's *Elements*, which they argue shows that knowledge of the relationship between the golden ratio and Fibonacci numbers was already established in the sixteenth century, as was the property of Fibonacci numbers described by Kepler. If true, this would predate Kepler by at least half a century.
7. Johannes Kepler, *The Six-Cornered Snowflake: A New Year's Gift*, trans. Jacques Bromberg (Philadelphia: Paul Dry Books, 2010), 67.
8. For more details on Kepler's odd ideas regarding sex and the golden section, see Van der Schoot, "Kepler's Search."
9. In the original German, the long title of Zeising's book begins with *Neue Lehre von den Proportionen des menschlichen Körpers*.

10. Albert van der Schoot, email message to Christophe Golé, July 28, 2021. In 2016, a German translation of Van der Schoot's book was published as *Die Geschichte des goldenen Schnitts* (History of the golden section).
11. Van der Schoot, email to Christophe Golé.
12. Van der Schoot, email to Christophe Golé.

## Notes to Chapter 4

1. Charles Bonnet, *Recherches sur l'usage des feuilles dans les plantes* . . . (Göttingen and Leiden: E. Luzac, fils, 1754), iii.
2. Bonnet, *Recherches sur l'usage*, vi–vii.
3. Charles Bonnet, *Collection complette [complète] des œuvres de Charles Bonnet* (Neuchâtel: Samuel Fauche, 1779), vol. 4, 10. The dew quotation is from Bonnet's "sketch" summarizing his work in 1765, which was added to this collected edition of his works.
4. Bonnet, *Recherches sur l'usage*, 160.
5. Bonnet, 53.
6. Julius von Sachs, whom we will meet again in chapter 17, in his *History of Botany*, was nothing short of snide in describing Bonnet's efforts: "It is impossible to imagine worse-devised experiments on vegetation . . . he ought certainly to have left the leaves on the living plants and observed the effect of the supposed absorption of dew on the vegetation. It is to be observed that by rising dew he evidently meant aqueous vapour, for the real dew descends chiefly on the upper side of the leaf; and what could he have expected to learn by laying cut leaves on water? How could this prove that leaves absorb dew? Nevertheless Bonnet came to the conclusion that the most important function of leaves was to absorb dew." See Julius von Sachs, *History of Botany, 1530–1860*, trans. Henry E. F. Garnsey, rev. Isaac Bayley Balfour (Oxford, UK: Clarendon Press, 1906), 487.
7. Bonnet, *Recherches sur l'usage*, 78.
8. Plants absorb the vast majority of their water through their roots. One could argue, in favor of Bonnet's thesis, that some plants, such as the potato, do have most of their stomata (breathing pores) on the undersides of their leaves and that some desert plants do get water from dew through their leaves. While water absorption through leaves' undersides is theoretically not impossible, it is not part of the modern list of leaves' essential functions (e.g., photosynthesis, transpiration, water storage, and so on).
9. Bonnet, *Recherches sur l'usage*, 24.
10. Bonnet, 24.
11. Bonnet, 52.
12. Bonnet, 166.
13. Bonnet, 159.
14. Bonnet, 160.
15. Bonnet, 160.
16. Beneath the plant showing redoubled spirals, Calandrini has drawn three seven-sided polygons (heptagons). Each is rotated 1/3 of a turn from the previous one, as he figures out the angular

positions of the leaves on the three helices. This 1/3 of a turn is then shown as a triangle. Bonnet also mentions plants with redoubled spirals organized in groups of 11, with five parallel spirals.

17. Bonnet, *Recherches sur l'usage*, 166.
18. Bonnet, 164.
19. Bonnet, 165.
20. In this diagram, we made choices that are not explicit in Bonnet's description but that are consistent with his more detailed description of the quincunx. For one, we raised the starting point of the yellow spiral by a third of the height between the first two points of the blue spiral. We also raised the starting point of the red spiral by twice that amount. Bonnet does not specify the pitch of the spirals. In our version, these 3 spirals are actually parastichies, and 5 parastichies can be discerned winding in the opposite direction. Slower pitches would yield parastichy numbers (8, 5), or (13, 8), as we will see with the Bravais brothers in chapter 7. While later scientists will present more systematic ways of describing these patterns, Bonnet laid the groundwork, and his approach is nonetheless compatible with our modern views.
21. Bonnet, *Recherches sur l'usage*, 165.
22. By seeing the quincunx's more vertical rows as spiraling along the stem rather than straight, Bonnet leaned toward the opinion of the Bravais brothers (see chapter 7), not the argument of Schimper and Braun (see chapters 5 and 6).
23. As quoted in Vincent Barras, "Histoire d'un syndrome 'Charles Bonnet,'" *Mémoires de la Société de physique et d'histoire naturelle de Genève* 47 (1994): 245–46. Bonnet's maternal grandfather, Charles Lullin, whose hallucinations Bonnet described, died in 1761 in his 92nd year. The quotation has been edited for clarity.
24. This, Bonnet's best-known quotation in English, is loosely translated from his *Essai analytique sur les facultés de l'âme* (Analytical essay on the faculties of the soul), (Copenhagen: Cl. and Ant. Philibert, 1760), 428.

## Notes to Chapter 5

1. There was a third friend in the group, Swiss scientist Alexander Agassiz. Schimper, Braun, and Agassiz were so often seen together that people called them "the cloverleaf." In recent years, Agassiz's racist idea that Black people were a separate human species has destroyed his reputation, and many institutions that honored him are being renamed.
2. Von Sachs, *History of Botany*, 163.
3. The earliest use of the word "phyllotaxis" that we have found appears in the 1830 expanded edition of Schimper's article "Beschreibung des Symphytum zeyheri . . . ," *Geigers Magazin für Pharmacie* 29, p. 46. The article was originally published *without* the word "phyllotaxis" in 1829, in the same journal. (Note that Mario Livio, the popular science author of *The Golden Ratio*, incorrectly credits Charles Bonnet with coining the word "phyllotaxis." In fact, Bonnet used the French term *arrangement des feuilles*.)
4. Karl Schimper, "Beschreibung des Symphytum zeyheri . . . ," *Geigers Magazin für Pharmacie* 29 (1830): 4. In German, Schimper's new "divergence angle" is *Divergenz*.

5. In the original German, Schimper's Ice Age sounds more poetic: *Eiszeit*.
6. Schimper also wrote a 50-page poem about love, money, and phyllotaxis: *Flieder und Goldlack, ein poetischer Brief über Zahlen und Dinge* (Lilac and wallflower, a poetical letter about numbers and things), published in 1847. See https://www.digitale-sammlungen.de/en/details/bsb10119426, starting on p. 251.
7. Schimper, "Beschreibung," 3.
8. This spiral, which generates all the visible spirals (parastichies), is called the generative spiral by Braun in his 1831 book on pinecones, on p. 221. Braun uses the word *Grundwendel*, "fundamental spiral." Schimper uses *Wendel* ("Beschreibung," 9).
9. Schimper calls these lines *Richtungs-Linien der Blätter*, "direction-lines of the leaves" ("Beschreibung," 5).
10. *Note from the future:* Given that 4 divides into 12, the numbers are not relatively prime. This means that in three steps, we return to the direction-line where we started (and we will not land on any direction-lines other than the three we visited). This is a graphic way to see whether a pair of numbers are relatively prime and obtain a reduced fraction.
11. Schimper, "Beschreibung," 9.
12. Remember that 3/8, for example, means that you make three turns around the stem as you count the eight leaves in the cycle. The ninth leaf is directly above the first.
13. Schimper, "Beschreibung," 25.
14. "Blumen machen mir Gesichter, / Jede weis und sagt etwas; / Aus den Blumen werden Lichter, / Und so lern' ich dies und das." Karl Schimper, *Gedichte* (Erlangen, Germany: Ferdinand Enke, 1840), 191. Translation by Nancy Pick.

## Notes to Chapter 6

1. The German title of Braun's 1831 book is *Vergleichende Untersuchung über die Ordnung der Schuppen an den Tannenzapfen als Einleitung zur Untersuchung der Blattstellung überhaupt* [Erlangen, etc., 1831].
2. For more on the strange breakdown of Schimper and Braun's friendship, see William M. Montgomery, "The Origins of the Spiral Theory of Phyllotaxis," *Journal of the History of Biology* 3, no. 2 (1970): 317–20.
3. Braun, *Vergleichende Untersuchung*, 10. Translation by the authors.
4. Cecilie Braun's pinecone drawings are held by the Botany Library of the Natural History Museum in London.
5. Stéphane poses two questions: Why stop at 21? And how do you determine verticality? Here we see that the direction of the 21st spiral starting from scale 1 oscillates a little from left to right. This is a good indication that on average it is close to vertical.
6. Braun finds that the generative spiral makes 8 turns around the cone between scales 1 and 22. From this, he deduces that the divergence angle of the generative spiral is 8/21 (fig. 6.5). This enables him to find the divergence angle of other spirals. For instance, the (yellow) spiral, passing through 27, 30, 33, . . . would have a divergence of $3 \times 8/21 = 24/21$, as it proceeds 3 steps at a time. But this corresponds to the smaller visible angle of $3/21 = 1/7$.

7. You can think of this diagram as having polar coordinates $(r, \theta)$, where $r$ represents the height in the cylindrical pinecone.
8. The Archimedean spiral, in polar coordinates $(r, \theta)$, can be described by the equation $r = a + b*\theta$ with real numbers $a$ and $b$. The coefficient $b$ is the "constant speed." (In fig. 6.7, you can see that the concentric circles are equidistant from one another.) You can draw one yourself by unwinding a thread wrapped around a pencil. Look for instructions online.
9. "But not quite" is important here, for leaf alignment. This is exactly what the Bravais brothers will argue is the flaw in considering exact fractions as divergence.
10. Montgomery, "Origins of the Spiral Theory," 299.

## Notes to Chapter 7

1. Auguste Bravais, *Notice des travaux scientifiques* (Paris: Mallet-Bachelier, 1854), 19. All translations in this chapter are by the authors.
2. Louis Bravais and Auguste Bravais, "Essai sur la disposition des feuilles curvisériées," *Annales des sciences naturelles*, 2nd ser., 7 (1837): 42–110.
3. Bravais and Bravais, "Essai sur la disposition," 42.
4. Bravais and Bravais, 42.
5. See Roger V. Jean, *Phyllotaxis: A Systemic Study in Plant Morphogenesis* (Cambridge: Cambridge University Press, 1994), 31–47.
6. It is also possible that Braun introduced this technique when drawing unrolled fir cones.
7. To mathematicians, this lattice is a representation of the group of integers $Z$ in the cylinder, itself a group $S^1 \times R$. The parastichies are the cosets of the quotient groups $Z/8Z$ and $Z/5Z$, of which there are respectively 8 and 5.
8. That is, without a common divisor other than the integer 1.
9. The Bravais brothers gave this the evocative name "encyclic number."
10. Here, Auguste's approach was not unlike that of Braun.
11. In the case of their figure 1, where $m = 5$ and $n = 8$; and $\Delta_m = 2$ and $\Delta_n = 3$, you take the 2 steps up from leaf 0 (to 8 and 16), and then 3 steps down (to 11, 6, and 1) to get to leaf 1.
12. For those who know continued fractions, if $m>n$, $\Delta_m/\Delta_n$ is the second to last convergent of the continued fraction of $m/n$.
13. Bravais and Bravais, "Essai sur la disposition," 64.
14. Bravais and Bravais, 70.
15. "A quantity is split in extreme and mean reason if it's divided in two unequal parts, the smallest being in the same ratio to the largest as the largest is to the whole." Bravais and Bravais, "Essai sur la disposition," 74.
16. If asked about the golden ratio, Auguste Bravais would have no doubt quickly produced the continued fraction for it, figuring that its successive truncations are quotients of adjacent Fibonacci numbers, thus proving what Kepler had stated a couple of centuries earlier.
17. The parastichy numbers must add up to this number.

## Notes to Chapter 8

1. Founded in 1807, the Friedrich Hofmeister music publishing house exists to this day.
2. Karl von Goebel, *Wilhelm Hofmeister: The Work and Life of a Nineteenth Century Botanist*, trans. H. M. Bower (London: Ray Society, 1926), 30.
3. Von Goebel, *Wilhelm Hofmeister*, 30.
4. Von Goebel, 81.
5. Wilhelm Hofmeister, *Allgemeine Morphologie der Gewächse* (General morphology of plants) (Leipzig: Engelmann, 1868).
6. Hofmeister, *Allgemeine Morphologie*, 85.
7. Incidentally, Nägeli later became notorious for discouraging Gregor Mendel from pursuing his work on genetics.
8. Von Goebel, *Wilhelm Hofmeister*, 86.
9. Von Goebel, 89–90. Edited for clarity.
10. Hofmeister notes that if you look closely at broccoli stems, for example, you can still see the original primordia contacts in scales linking the leaflets.
11. Donald R. Kaplan and Todd J. Cooke, "The Genius of Wilhelm Hofmeister: The Origin of Causal-Analytical Research in Plant Development," *American Journal of Botany* 83, no. 12 (1996): 1647.

## Notes to Chapter 9

1. As quoted in A. Zimmermann, "Simon Schwendener," *Berichte der Deutschen Botanischen Gesellschaft* 40, no. 11 (1922): 59.
2. This highly mathematical microscope handbook discussed such topics as diffraction, polarization, objectives, and eyepieces, becoming Schwendener's only book translated into English.
3. Anja Geitmann, Karl Niklas, and Thomas Speck, "Plant Biomechanics in the 21st Century," *Journal of Experimental Botany* 70, no. 14 (2019): 3435.
4. Simon Schwendener, *Das mechanische Prinzip im anatomischen Bau der Monocotyledonen* . . . (The mechanical principle of the anatomy of monocotyledons) (Leipzig: Engelmann, 1874).
5. Simon Schwendener, *Die periodischen Erscheinungen der Natur insbesondere der Pflanzenwelt* (Zurich: Höhr, 1856), 1. In the original German: "Alle organische Leben beruht auf den Spiele forwährender Bildung und Entbildung."
6. Rosemarie Honegger, "Simon Schwendener (1829–1919) and the Dual Hypothesis of Lichens," *Bryologist* 103, no. 2 (2000): 312.
7. Honegger, "Simon Schwendener."
8. Zimmermann, "Simon Schwendener," 63. Translation by the authors.
9. Their discoveries were linked to technological advances, as Zeiss and other German optics manufacturers made major improvements in microscope resolution and illumination.
10. Schwendener, *Das mechanische Prinzip*, 46. Translation by the authors.
11. Schwendener, 57. The precise laying out of assumptions is very much a modern practice in scientific modeling.
12. Chris notes: "Disk packing" or "disk stacking"? And what about "circle packing?" Disk stacking is the term that Stéphane and I used in the paper we published on convergence. The term "circle

packing"—which stems from Kepler's packing of pomegranate seeds and cannon balls—has come to mean "an arrangement of circles inside a given boundary such that no two overlap and some (or all) of them are mutually tangent." The circle packing community is all about optimization of packing density, and not about any dynamical process that would build one circle at a time. So circle packing is like lattices, contrary to our dynamical view. All this is to say that while the term "disk stacking" is not ideal and does not yet have widespread usage, I think that either "circle packing" or "disk packing" might be misleading.

13. Schwendener indicates the three whorls by I, I′, I″, then II, II′, II″, and finally III, III′, III″. The first row is made up of bracts protecting the flower bud, and the next two rows are the flower petals. After the Fibonacci patterns come the pistils.
14. Schwendener, *Das mechanische Prinzip*, 11. Translation by the authors.
15. This modeling concept was a first in botany, to our knowledge.
16. In figure 9.11, the force diagram that Schwendener labels his figure 2, which attempts to give a physical interpretation of vertical pressure, is somewhat flawed. Schwendener wrongly claims that the lateral forces $\overline{am}$ and $\overline{bn}$ have the same magnitude, as noted by Irving Adler, Denis Barabé, and R. V. Jean in "A History of the Study of Phyllotaxis," *Annals of Botany* 80, no. 3 (1997): 235.
17. It is important to note that, contrary to the perception of Schwendener's work by some scientists who cite him, he did not see pressure as the *cause* of the deformation, but instead as equivalent to when the stem grows faster in a lateral direction than a longitudinal one.

## Notes to Chapter 10

1. Adrianus D. J. Meeuse, "G. van Iterson Jr.," *Acta botanica neerlandica* 18, no. 1 (1969): 10.
2. Gerrit van Iterson, *Mathematische und mikroskopisch-anatomische Studien über Blattstellungen* (Jena, Germany: Fischer, 1907).
3. "Professor G. van Iterson," *Nature* 237 (1972): 357.
4. Meeuse, "G. van Iterson Jr.," 10. Punctuation edited for clarity.
5. "Professor G. van Iterson," 358.
6. Gerrit van Iterson, *Nieuwe Studiën over bladstanden I* (Netherlands: N. V. Noord-Hollandsche Uitgevers Maatschappij, 1964).
7. The simple arcs of a circle appear in Van Iterson's diagram when plotted in the plane where the x-axis represents the divergence, and the y-axis the height of leaf 1, with leaf 0 at the origin. Van Iterson painstakingly produced plots in several coordinate systems.
8. The symmetric case of (2, 1) is not shown.
9. Those familiar with hyperbolic geometry will recognize these arcs of circles as hyperbolic geodesics in the Poincaré upper half-plane model. Moreover, each of these circles is the image of the y-axis (the "arc of circle" between 0 and infinity) by a hyperbolic isometry in the group $PSL(2, \mathbb{Z})$ of integer Möbius transformations with determinant 1. The fact that these transformations preserve angles explains the similarity of branching angles in the diagram. Note, however, that it is extremely unlikely that Van Iterson made this connection. For more on this, see Pau Atela, Christophe Golé, and Scott Hotton, "A Dynamical System for Plant Pattern Formation," *Journal of Nonlinear Science* 12 (2003): 641–76.

10. Hexagonal lattices show 3 sets of parastichies. For example, the lattice corresponding to the meeting of the (2, 3) and (3, 5) branches has 2 parastichies going one direction, 3 going in the opposite direction, and 5 more vertical ones.
11. Note that Van Iterson's branch also matches Schwendener's zigzagging arcs from the last chapter, in figure 9.11. In addition to his work on the Fibonacci branch, Van Iterson derived the equations for the arcs' circles using mathematical tools introduced by the Bravais brothers, but applied more systematically.
12. Van Iterson, *Mathematische*, 259n1.
13. Van Iterson, 259n1.
14. Van Iterson, 260.
15. Van Iterson, 260.
16. A critical review of Van Iterson's book in *Nature* stated: "Iterson's volume including 300 pages on the botanical aspect of the question, devotes 190 to mathematical speculations, the greater part of which will therefore not appeal to the average botanist at all." *Nature* 77, no. 1990 (1907): 145.
17. Van Iterson was perhaps too eager to embrace the work of Church, a leading phyllotaxis authority at the time. Church claimed that parastichies crossed at right angles, and he considered only transitions between *zickzacklinie* that met at a right angle. Unfortunately, this is not quite botanically or mathematically realistic.
18. Lesley A. Robertson, Delft University of Technology, email message to Nancy Pick, July 10, 2021.
19. In an email message to Chris, mathematician-author Jonathan Swinton noted that he has nonetheless found Church's phyllotaxis work useful, particularly his compilation of parastichy numbers for many plants.
20. Arthur Harry Church, *On the Relation of Phyllotaxis to Mechanical Laws* (London: Williams and Norgate, 1904), 52.
21. Church, *On the Relation of Phyllotaxis*, 52.

## Notes to Chapter 11

1. Andrew Hodges, *Alan Turing: The Enigma* (Princeton, NJ: Princeton University Press, 2012), 207–8. The first edition was published in 1983.
2. In 1950, Turing published his seminal paper "Computer Machinery and Intelligence," introducing the idea of artificial intelligence.
3. Turing also knew about biological patterns from D'Arcy Wentworth Thompson's *On Growth and Form*, a familiar reference for many scientists in this book.
4. Alan Turing, "The Chemical Basis of Morphogenesis," *Philosophical Transactions of the Royal Society B* 237, no. 641 (1952): 37–72.
5. This idea, how the unfolding of simple dynamics can lead to complex results, is the possible link between Hofmeister's basic observation of primordia appearing one by one in the largest space—visible in his drawings—and Fibonacci's mathematical ideal. It follows that a simple arithmetic computation can lead to such developments as artificial intelligence.
6. Turing's letter to Harold Scott MacDonald "Donald" Coxeter was dated May 28, 1953, as quoted in Coxeter, "The Role of Intermediate Convergents in Tait's Explanation for Phyllotaxis," *Journal of Algebra* 20 (1972): 167.

7. Jonathan Swinton's paper appears as a chapter in S. Barry Cooper and Jan van Leeuwen, eds., *Alan Turing: His Work and Impact* (Amsterdam: Elsevier, 2013). Quote is on p. 834.
8. Jonathan Swinton, *Alan Turing's Manchester* (Cheltenham, UK: History Press, 2022), 171.
9. This theoretical device is now called a Turing machine.
10. Turing saw that as the growing plant's phyllotaxis travels down the (3, 5) branch of the Van Iterson diagram, it arrives at a fork where it meets the (3, 8) and (5, 8) branches. Continuing its growth, the plant almost always chooses the (5, 8) branch. In general, traveling down the ($m$, $n$) branch (recall that $m < n$), the plant will choose the branch ($m$, $m + n$) at the fork, and not ($n$, $m + n$). Turing called that the fundamental hypothesis of phyllotaxis. (Note that this is entirely different from the fundamental *theorem* of phyllotaxis originally seen by the Bravais brothers in chapter 14.)
11. Swinton, *Alan Turing's Manchester*, 171.
12. Alan Turing, *Morphogenesis*, ed. P. T. Saunders, vol. 3, *Collected Works of A. M. Turing* (Amsterdam: North-Holland, 1992).
13. Jonathan Swinton, Erinma Ochu, and the MSI Turing's Sunflower Consortium, "Novel Fibonacci and Non-Fibonacci Structure in the Sunflower: Results of a Citizen Science Experiment," *Royal Society Open Science* 3, no. 5 (2016): 160091.

## Notes to Chapter 12

1. Ian Stewart, "Daisy, Daisy, Give Me Your Answer, Do," *Scientific American* 272, no. 1 (1995): 99.
2. L-systems are now extensively used for simulations of space occupation and interaction of plants. Initially, Lindenmayer worked with yeast, filamentous fungi, and bacteria. A few years later, because L-systems are particularly good at describing growing branching patterns, they would also be used for fractals, which were first formally described in 1975.
3. These cell-to-cell interactions would become famous when simplified in the "Game of Life," invented in 1970 by British mathematician John Horton Conway.
4. The inhibitor, which was unspecified, could be an enzyme or other chemical.
5. In 1975, Lindenmayer coauthored a paper using the same computer model to simulate transitions in the irregular phyllotaxis of the succulent *Bryophyllum tubiflorum*, but from a (2, 2) whorled mode to (3, 3). (The plant, native to Madagascar, has been renamed *Kalanchoe delagoensis*.)
6. See Arthur H. Veen and Aristid Lindenmayer, "Diffusion Mechanism for Phyllotaxis," *Plant Physiology* 60, no. 1 (1977): 132 (table 1).
7. This simulation shows a (1, 2) lattice.
8. The cells in the new layer have average inhibitor concentration.
9. Chris notes: In 1975, John H. M. Thornley, who acknowledged the influence of Turing's "Chemical Basis of Morphogenesis" via Claude Wardlaw, introduced a model of diffusion and decay of just one morphogen. For simplicity, he assumed that diffusion occurred only along the periphery of the meristem. While the effect of the 2D geometry of the plant surface was lost, this approach had the advantage of making the math more tractable. He too looked for patterns of constant divergence angle—which he found. He then published a paper relating divergence angles and

plastochrons to parastichy numbers. Also in the mid-1970s, Graeme J. Mitchison used the computer to study phyllotaxis. But his main goal was to give a heuristic argument for what Turing called his hypothesis of geometric phyllotaxis.

10. He wondered which lattice would result in the minimal interaction energy. A real system would go toward this minimal interaction: the energy remaining above the minimum pushes it toward this minimum.

11. The symmetry is by a group $PSL(2, Z)$ of complex transformations of the form $z \rightarrow az + bcz + e$, where $z = d + ir$ is the point $(d, r)$ seen as a complex number and $a, b, c,$ and $e$ are integers satisfying $ae - bc = 1$. Do you recall this relationship?

## Notes to Chapter 13

1. Philippe Édouard Léon van Tieghem, *Traité de botanique* (Paris: Savy, 1884).
2. Lucien Plantefol, *La théorie des hélices foliaires multiples* (Paris: Masson, 1948).
3. Conditions would be similar to the repulsion found in Levitov's vortices, except that this time the drops would be added one by one, not assumed to be already on a lattice.
4. Stewart, "Daisy, Daisy," 96.
5. The laws of symmetry determine that each turn moves in the opposite direction from the previous one: at the bifurcation, the flatter contact parastichies are replaced by the new diagonal parastichies, which are now the steepest. So with each new step, the same case repeats on the other side.
6. Stewart, "Daisy, Daisy," 99.
7. Before that point on the Van Iterson arc, the two parastichies would run in the same direction, which is forbidden (because the new disk would not be in a minimum between two previous ones).
8. Stéphane's formula was simple and valid for each branch, using only the limit values for which it would be hexagonal. As Chris pointed out to Stéphane, a single formula could be valid for different branches because of the self-similarity of the tree, allowing renormalization.
9. These dashed lines also start on the x-axis on Levitov's "mountain summits," where the divergence is rational.

## Notes to Chapter 14

1. The model used by Pau Atela, Scott Hotton, and Chris was inspired by the Hofmeister model of Stéphane and Yves Couder: primordia appear one at a time, like the droplets of ferrofluid in their experiment. The three coauthors pushed this model to its limit, where the interaction between primordia decreases so quickly with distance that, in effect, the new primordium "feels" only its nearest neighbor. With that assumption, they showed that stationary solutions were stable. Using the (hyperbolic) symmetries of the Van Iterson diagram, they showed that the stationary solutions were lattices *on* the Van Iterson diagram, not just close to it.
2. In fact, Chris was using disk stacking rules that Schwendener had established more than 130 years before, but he had not yet discovered Schwendener's work. In developing his model, Chris based it on another model by Stéphane and Yves Couder, which they called their Snow

and Snow model: it assumes that primordia form where and *when* there is enough space. Chris had pushed that model to its limit as in the first paper with Atela and Hotton, assuming steep decay with distance of the interaction between primordia.

3. In figure 14.3, rhombic tilings show their periodicity geometrically: the angles that parastichies form at a disk repeat every sixth disk, such that the angles at disk 6 are the same as at disk 12. Likewise, the angles at disks 4 and 10 are the same. Note that $6 = 3 \times 2$, the product of the parastichy numbers. More generally, in a rhombic tiling of parastichy numbers $(m, n)$, the period is $mn$. The disk lattices of Schwendener and Van Iterson, also called rhombic lattices, are a special case of rhombic tilings, with straight parastichies. Chris's experiments were evidence that patterns always converge to rhombic tilings in the disk stacking process with fixed disk size, but not necessarily to regular lattices. It took many years for Chris and Stéphane to rigorously determine the mechanism of convergence. In the language of dynamical systems, rhombic tilings form the attractor in the stacking process. Interestingly, they found that a pattern could sometimes become a rhombic tiling after a finite number of disks were stacked, yet a similar pattern could take infinitely many disks to converge to a rhombic tiling.

4. Some time later, Stéphane pointed out to Chris that Van Iterson had already written about a *zickzacklinie*, or zigzag line (see chapter 10). Although Van Iterson used zigzag lines in a more limited way than Chris and Stéphane use fronts, this is yet another example of how science is self-repeating even as it moves forward!

5. As for the Fibonacci pattern along branches and other organs, there is some evidence that the main stem and basal leaf may serve the same role as the cotyledon, initiating a (1, 1) or (2, 2) pattern that then becomes other Fibonacci spirals.

6. Stéphane first encountered the special case of corn when an audience member asked a question at a public lecture: Why do corn kernels always display even numbers of nearly vertical columns? Back in the 1700s, Charles Bonnet had wondered the same. Stéphane found that the columns on the corncob have their origins in the nearly vertical diagonals: they are vertical when the two parastichy numbers are equal, and they are slightly inclined when the parastichy numbers differ by 1. They are even because each primordium produces two side kernels, doubling the number of vertical diagonals.

7. Alexander Braun had already described *Banksia* phyllotaxis, but Stéphane didn't know it at the time.

8. In 2016, Stéphane and Chris started writing an article about the duality of Fibonacci/quasi-symmetric phyllotaxis, and they invited Jacques to join in. This was an eventful few months for the three of them: Jacques and Chris both lost their mothers in the same week, and a month later, Stéphane's second child was born. But the three thought that the best thing for everyone, including the baby, was to soldier on with the work. Chris felt, as he worked next to his unconscious mom in the days preceding her death, that she would have wanted him to do exactly that.

## Notes to Chapter 15

1. Georg Cantor, *Gesammelte Abhandlungen* (Berlin: Springer, 1932), 374. English translation by Rudy Rucker, *Infinity and the Mind* (Princeton, NJ: Princeton University Press, 2005), 43.

2. This is yet another example of Stigler's law of eponymy, where the name attached to a scientific discovery is not that of the first person to describe it: the "Cantor set" was discovered by Henry John Stephen Smith nine years earlier, in 1874. Nevertheless, Cantor did study this set in considerable depth.
3. A note here for the mathematically inclined. One of Cantor's greatest discoveries is that for subsets of real numbers, there can be two types of infinities: countable (like the natural numbers 0, 1, 2, . . . or the set of fractions between 0 and 1), and uncountable (like the whole real line, or any interval in it). The Cantor set is a beautiful example of a set that contains no interval yet is uncountable! A question that kept Cantor awake at night still remains to be proven or disproven: whether there is another kind of infinity between the countable and uncountable (the continuum hypothesis).
4. Like the Cantor set, the Koch snowflake is a treasure chest of counterintuitive properties: it has, for example, an infinite perimeter yet encloses a finite surface. It is also a continuous curve with no tangent line at any of its points. See Šime Ungar, "The Koch Curve: A Geometric Proof," *American Mathematical Monthly* 114, no. 1 (2007): 61–66.
5. Benoit Mandelbrot gave himself the middle initial B., which did not stand for anything!
6. Published in *Science*, Mandelbrot's 1967 paper "How Long Is the Coast of Britain?" was inspired by the work of Lewis Fry Richardson, a British mathematician, meteorologist, and pacifist who—thinking that war was above all a question of borders between countries—tried to define them and came up with this problem of varying border length.
7. Nigel Lesmoir-Gordon, "Benoît Mandelbrot Obituary," *The Guardian*, October 17, 2010. (Note that Mandelbrot dropped the accent in his first name.)
8. Mandelbrot required that a fractal have a greater Hausdorff dimension (a measure of roughness) than topological dimension (the latter being an integer). A Hausdorff dimension is a precise way to determine this exponent, which lies between 0 and 2 in the plane. Although most of the time this implies that the Hausdorff dimension of the set is not an integer, the boundary of the Mandelbrot set has a topological dimension of 1 and a maximum Hausdorff dimension of 2.
9. In a 1999 paper, Polish biologist Wojciech Borkowski analyzed the leaves of many plants for their fractal structure. He concluded that "leaf outlines are not fractals in the strict sense. Most leaves are not self-similar in terms of fractal geometry. For example, semi-self-similar leaves include the leaf of the common chervil but not the maple leaf. Leaf outlines as fractal objects are not usually scale invariant. This means that a leaf may exhibit fractal features for some scale ranges only. Moreover, these features can differ with the scale range." Wojciech Borkowski, "Fractal Dimension Based Features Are Useful Descriptors of Leaf Complexity and Shape," *Canadian Journal of Forest Research* 29, no. 9 (1999): 1304.
10. Technically speaking, the shaft of a compound leaf is called a rachis.
11. In a beautiful interaction between mathematical and computational modeling (using L-systems in particular), as well as gene editing, Christophe Godin, François Parcy, and their teams were able to make *Arabidopsis*, a cousin of Romanesco broccoli, produce fractal curds. Instead of completing the flowering stage, the meristem in these mutants grows new shoots that "remember" that they have already gone through the flowering stage. These shoots themselves reiterate

the process. See Eugenio Azpeitia et al., "Cauliflower Fractal Forms Arise from Perturbations of Floral Gene Networks," *Science* 373, no. 6551 (2021): 192–97. For a less technical view of this work, see Sabrina Imbler, "Cauliflower and Chaos, Fractals in Every Floret," *New York Times*, July 8, 2021.

12. In Kenneth Falconer's *Fractal Geometry*, he lists some properties of fractals (slightly edited here) as "too irregular to be described in traditional geometric language, both locally and globally"; "often has some form of self-similarity"; "usually, the 'fractal dimension' is greater than its topological dimension"; "in most cases of interest [the set] is defined in a very simple way, perhaps recursively." Kenneth Falconer, *Fractal Geometry* (Hoboken, NJ: John Wiley and Sons, 1990), xxv.

## Notes to Chapter 16

1. Unfortunately for Charles Bonnet, whom we met in chapter 4, these veins play no role in pumping dew.
2. The crucial point was observing how the leaf lamina extends all the way to the edge of the folded shape. When the leaf was folded inside the bud, the curved edge that delimited its growth was always very clear. Once you know how the leaf was folded and the shape of the edge, you can then re-create the *unfolded* leaf.
3. By trying to quantify a wide range of plant characteristics, in a sense Michel Adanson anticipated the current approach to building a biological family tree, based in part on the similarity of several pieces of DNA.

## Notes to Chapter 17

1. Julius von Sachs, *Lectures on the Physiology of Plants*, trans. H. Marshall Ward (Oxford, UK: Clarendon Press, 1887), 499.
2. Von Sachs, *Lectures on the Physiology of Plants*, 500n.

## Notes to Chapter 18

1. Over the course of writing this book, the authors often discussed the nonlinear advancement of discoveries, a concept they called the "spiral of science." Ironically, that phrase itself had been used before. See Luke A. Schwerdtfeger, "Spirals of Science," *Science* 362, no. 6420 (2018): 1318. A spiral keeps turning without end—just as in science, Jacques noted, "the questions are never really answered, they are simply reformulated with more details and greater clarity." In addition, a spiral always arcs back on itself. In just this way, he said, "scientists constantly revisit earlier works, often 'rediscovering' for themselves observations that had already been made."
2. Léo Errera, "Sur une condition fondamentale d'équilibre des cellules vivantes," *Comptes rendus hebdomadaires des séances de l'Académie des sciences* 103 (1886): 822. Translation by Jacques Dumais.
3. D'Arcy Wentworth Thompson, *On Growth and Form*, rev. ed. (Mineola, NY: Dover, 1992), 568 (italics added).

4. While the cell patterns appear particularly clearly in the Venus flytrap trichomes, Jacques noted, they're actually quite common. Basil and hemp also produce pungent essential oils on their hairlike trichomes, to discourage animals that brush against them from eating the plant. Their cells also provide good examples of Errera's rule, as implemented by Thompson.
5. Norbert Wiener, "The Shortest Line Dividing an Area in a Given Ratio," *Proceedings of the Cambridge Philosophical Society* 18 (1916): 56.
6. Richard P. Grant, "Plant Cells and Soap Bubbles," *The Scientist*, August 2011.
7. "There were clues in the literature," Jacques said. "If we had found these earlier, we could have moved forward faster. For example, in 1980, T. J. Cooke had shown that in a fern, you could go back and forth between different cell divisions, and the proportion might depend on cell length. But he never looked at other species beyond the fern he was studying."
8. The cell's predilection for economy explains why Thompson's categorical version of Errera's rule has had some success in predicting division patterns in plants, even if less frequent configurations are ultimately lost in Thompson's formulation.
9. Errera's student in Belgium, Émile Auguste Joseph De Wildeman, also explicitly stated in his 1893 paper that plants behaved like bubbles, with more than one possibility for where they divided. Émile De Wildeman, "Études sur l'attache des cloisons cellulaires," *Mémoires couronnés . . . , Académie royale des sciences, des lettres et des beaux-arts de Belgique* 53 (1893): 1–84.
10. Connecting the dots here, Etienne Couturier, whom we met in chapter 16 in the discussion of leaf folding, worked as a postdoc with Jacques on this diagram.
11. Note that phyllotaxis occurs at the periphery of the meristem, not at the center.

## Notes to Chapter 19

1. Hugh Falconer, "On the American Fossil Elephant of the Regions Bordering the Gulf of Mexico . . . ," *Natural History Review*, 1863, 80.
2. See Nicolas Di-Poï and Michel C. Milinkovitch, "The Anatomical Placode in Reptile Scale Morphogenesis Indicates Shared Ancestry among Skin Appendages in Amniotes," *Science Advances* 2, no. 6 (2016), e1600708; and Danielle Dhouailly, "A New Scenario for the Evolutionary Origin of Hair, Feather and Avian Scales," *Journal of Anatomy* 214, no. 4 (2009): 587–606.
3. The answer is no.
4. Justin P. Kumar, "Building an Ommatidium One Cell at a Time," *Developmental Dynamics* 241, no. 1 (2012): 137.
5. The transition from hexagonal to square patterning can also be seen in the three images of figure 14.11.
6. A dislocation in a plane hexagonal pattern is composed of a joint pair of pentagonal and heptagonal cells.
7. Twelve isolated pentagons are necessary to create a full spherical object, whatever the number of hexagons in between (this is a topological result).
8. A true logarithmic spiral would keep spiraling to an infinitely small center, which does not exist in nature. The Fibonacci construction, which starts with 1, is much closer to the logarithmic spirals found in mollusk shells, which begin with a small but finite larval size.

9. For every mollusk species, the shell displays a constant growth rate, as in a true logarithmic spiral.
10. The growth rate for a spiral built with Fibonacci numbers approaches the golden mean for 1/4 of a turn—or the golden mean to the fourth power for one full turn—which is approximately 6.854.
11. Charles Darwin to Asa Gray, April 20, 1863, Darwin Correspondence Project, "Letter no. 4110," https://www.darwinproject.ac.uk/letter/?docId=letters/DCP-LETT-4110.xml.
12. Darwin to Gray, May 11, 1863, Darwin Correspondence Project, "Letter no. 4153," https://www.darwinproject.ac.uk/letter/?docId=letters/DCP-LETT-4153.xml.
13. Gray to Darwin, May 26, 1863, Darwin Correspondence Project, "Letter no. 4186," https://www.darwinproject.ac.uk/letter/?docId=letters/DCP-LETT-4186.xml.
14. Gray to Darwin, June 10–16, 1863, Darwin Correspondence Project, "Letter no. 4198," https://www.darwinproject.ac.uk/letter/?docId=letters/DCP-LETT-4198.xml.
15. The idea that Fibonacci phyllotaxis maximizes light exposure on leaves recalls Charles Bonnet's mistaken notion, back in the eighteenth century, that phyllotaxis maximizes the ability of leaves to collect dew on their undersides.
16. Darwin seems to have embraced the concept that Fibonacci spirals might be optimal for young plant organs. Writing to J. D. Hooker on June 2, 1876, he states: "I cannot get it out of my head that [mathematician George Biddell Airy] is on the right course in explaining phyllotaxis by the mutual pressure of very young buds." Although it is true that Fibonacci patterns can be optimal in this way (Airy's idea being a precursor to Simon Schwendener's contact pressure concept), scant scientific support has emerged for this or any other functional explanation.

## Notes to Appendix

1. Leonardo da Vinci, *Literary Works*, vol. 1, 207 (section 402). Edited here for clarity.
2. In some old trees these veins can be directly seen as undulations of the bark. It is still debated why they spiral mostly in the right-hand direction, as drawn by Leonardo. Some scientists have proposed that it boils down to the chirality of the cellulose that composes the plant's cell walls.
3. Leonardo da Vinci, as quoted in William R. Thayer, "Leonardo da Vinci as a Pioneer in Science," *The Monist* 4, no. 4 (1894): 520. Also, the French essayist Montaigne recounted how, while traveling in Pisa in 1581, he heard a very similar account of tree rings from a jeweler known for his ingenious mathematical instruments. Had news of Leonardo's discovery already been making the rounds? See Alfred W. Bennett, "Leonardo da Vinci as a Botanist," *Nature* 2 (1870): 42–43.
4. Leonardo da Vinci, *Literary Works*, vol. 1, 205 (section 394). The source is Leonardo's Notebook 1, verso 12, held by the Institut de France. The two paragraphs quoted here are written not only in Leonardo's usual mirror writing, but also upside down.
5. Christophe Eloy, "Leonardo's Rule, Self-Similarity, and Wind-Induced Stresses in Trees," *Physical Review Letters* 107, no. 25 (2011): 258101.
6. See Royal Collection Trust, Leonardo da Vinci, "A Sheet of Miscellaneous Studies," recto, RCIN 912283.
7. Walter Isaacson, *Leonardo da Vinci* (New York: Simon and Schuster, 2017), 174.

8. Bonnet, *Recherches sur l'usage*, 177. All Bonnet translations are by the authors.
9. Note that 7 belongs to the Lucas sequence 1, 3, 4, 7, 11, . . . , the third most common sequence found in plants, after Fibonacci and double Fibonacci (2, 2, 4, 6, 10, . . .).
10. Was Bonnet blinded to this somewhat obvious observation by his admiration and respect for Calandrini, who first identified the five orders?
11. The spirals that Bonnet discussed here ("orthostichies") are not the most obvious, but they closely follow the axis of the stem. Whether these orthostichies are actually spiraling will be a point of contention for the next generation of phyllotaxis scientists. To check variation in spiral direction, Bonnet looked at 75 chicory plants. He counted 43 with spirals winding up the stem from right to left, and two with multiple stems winding in different directions. Bonnet also searched the other four orders for similar spiraling of vertically aligned leaves but did not find them.
12. Bonnet, *Recherches sur l'usage*, 180.
13. Bonnet, 175.
14. Bonnet sometimes called it *Mays*, but more often *Bled de Turquie*, or Turkish wheat.
15. Cauchy's demonstration eventually appeared in his *Exercises de mathématiques* from 1826, a compilation of his articles.
16. Moreover, Brocot's goal was very similar to Haros's in 1802: given a decimal value for a given gear ratio, one could find the number of teeth in each connecting gear coming close to this value at a given precision (the denominator). In practice, the ratio had to be drawn in the exact position along the interval, as Braun did, and limited to a given denominator, as Haros did. Haros also had the reverse aim: if someone had a fraction in mind (still using the English system), then it was easy to find the closest decimal value. Stern was more interested in the general question of generating fractions, so the successive lines were then simply ordered by the number of mediant operations.
17. To be exact, we are referring here to Poincaré's upper half-plane model of hyperbolic geometry.

# ILLUSTRATION CREDITS

The full-page photographs accompanying the title page and section divisions are used by kind permission of Victor Mozqueda. The plants are: *Eriosyce* sp. (p. 2); *Osteospermum ecklonis* (p. 18); *Sempervivum* sp. (p. 50); *Mammillaria magnifica* (p. 116); *Echinocereus reichenbachii* (p. 158); *Lithops* sp. (p. 244); and *Fragaria* × *ananassa* (p. 274). For the food photographs, his assistant was Samantha Miller.

In addition to taking the photos credited below, photographer Stephen Petegorsky made expert copies of many illustrations for this book.

Stéphane Douady (fig. 0.1). Sam, "Spiral Aloe from above," 2015, via Wikimedia Commons, CC BY-SA 4.0 (fig. 0.2). Redux of fig. 0.1, detail (fig. 0.3). Pau Atela, Christophe Golé, Michael Marcotrigiano, and Madelaine Zadik, based on Smith College exhibition "Plant Spirals: Beauty You Can Count On," annotated (figs. 0.4, 0.6). Left, Christophe Godin; right, S. Douady (fig. 0.5). C. Golé (fig. 0.7). SEM image used by kind permission of Rudolf Rutishauser (fig. 0.8). C. Golé (figs. 0.9, 0.10). Victor Mozqueda (fig. 0.11). S. Douady (figs. 0.12–0.14).

Cynthia J. Huffman, "Numeric hieroglyphs in the Louvre," 2022, via Convergence, CC BY 4.0 (fig. 1.1). Pierre-Joseph Redouté, "Common myrtle," via Rawpixel, CC BY 4.0 (fig. 1.2). S. Douady (fig. 1.3). Fibonacci, *Liber abaci*, 14th century ed., via Wikimedia Commons (fig. 1.4). Romain, "Fibonacci Spiral," 2022, via Wikimedia Commons, CC BY-SA 4.0 (fig. 1.5). Photos by Victor Mozqueda (fig. 1.6).

All Leonardo da Vinci images in chapter 2 have been flipped for legibility. Leonardo da Vinci, Notebook G, 33r, used by permission of the Bibliothèque de l'Institut de France (fig. 2.1). Details of fig. 2.1 (figs. 2.2–2.5). S. Douady (fig. 2.6).

Frederick Mackenzie, "Johannes Kepler engraving," via Wikimedia Commons (fig. 3.1). E. Benassit and E. Yon, "The hangman of Stuttgart shows Kepler's mother instruments of torture," via Wikimedia Commons (fig. 3.2). Victor Mozqueda (fig. 3.3). Leonardo da Vinci in Luca Pacioli's *Divina proportione*, plates 22 and 23, via Internet Archive (fig. 3.4). C. Golé et al., based on Smith College exhibition "Plant Spirals" (figs. 3.5, 3.6).

James Caldwell engraving, "Charles Bonnet," via the Wellcome Collection (fig. 4.1). Jan Wandelaar engraving in Bonnet, *Recherches sur l'usage des feuilles dans les plantes . . .* , plate

2 (fig. 4.2). Top, Bonnet, plate 20; bottom, S. Douady (fig. 4.3). Bonnet, 167 (fig. 4.4). C. Golé (fig. 4.5). Left, Bonnet, plate 21; right, S. Douady (fig. 4.6). C. Golé (fig. 4.7). Redux of fig. 4.5, left (fig. 4.8). C. Golé (fig. 4.9).

Karl Schimper portrait in *Allgemeine Illustrirte Zeitung*, no. 23, p. 365 (fig. 5.1). S. Douady (figs. 5.2–5.4). Schimper, "Beschreibung des Symphytum zeyheri . . . ," *Geiger's Magazin für Pharmacie* 29, p. 20 (fig. 5.5). Schimper, table 1, annotated by S. Douady (fig. 5.6). Left, Schimper, 27; right, Schimper, 28 (fig. 5.7).

Cecilie (Cécile) Braun self-portrait, reproduced by kind permission of the Ernst Mayr Library and Archives of the Museum of Comparative Zoology, Harvard University (fig. 6.1). All pinecone drawings in chapter 6 by Cecilie Braun, published in Alexander Braun, *Vergleichende Untersuchung über die Ordnung der Schuppen . . .* . Braun, plate 18 (fig. 6.2). Braun, plates 17 and 25 (fig. 6.3). Braun, plate 19 (figs. 6.4, 6.5). Braun, plate 50, annotated by S. Douady (fig. 6.6). Braun, plate 21 (fig. 6.7). Braun, plate 30 (fig. 6.8). Left, Braun, plate 28; right, S. Douady (fig. 6.9). Photo by Stephen Petegorsky of 1876 Braun poster, personal collection of C. Golé (fig. 6.10). S. Douady (figs. 6.11–6.13).

J. M. J. Bouillat, "Auguste Bravais, Voyageur et savant (1811–1863)," *Les contemporains*, no. 588, no. 1 (fig. 7.1). Auguste Mayer, "Corvette *La Recherche* near Bear Island on August 7, 1838," via Old Book Illustrations (fig. 7.2). Louis Bravais and Auguste Bravais, "Essai sur la disposition des feuilles curvisériées," *Annales des sciences naturelles*, 2nd ser., 7, plate 2 (fig. 7.3). C. Golé (fig. 7.4). Bravais and Bravais, plate 2, detail (fig. 7.5). Franck Le Driant, "Asphodeline lutea," 2016, via FloreAlpes, used by kind permission (fig. 7.6). Bravais and Bravais, 65 (fig. 7.7). Auguste Bravais, *Mémoire sur les systèmes formés par des points distribués régulièrement . . .* , plate 1 (fig. 7.8). C. Golé (fig. 7.9). C. Golé et al., based on Smith College exhibition "Plant Spirals" (fig. 7.10, no captions).

Left, Wilhelm Hofmeister, *Allgemeine Morphologie der Gewächse*, 498; right, S. Douady (fig. 8.1). Eduard Schultze, photo of Wilhelm Hofmeister, via TOBIAS-bild, Universitätsbibliothek Tübingen (fig. 8.2). Hofmeister, 493 (figs. 8.3, 8.4). Hofmeister, 459 (fig. 8.5). Left, SEM image used by kind permission of Rolf Rutishauser; right, C. Golé et al., based on Smith College exhibition "Plant Spirals" (fig. 8.6). Left, Hofmeister, 434; center and right, Hofmeister, 435 (fig. 8.7). Hofmeister, 523 (fig. 8.8).

Simon Schwendener, *Das mechanische Prinzip im anatomischen Bau der Monocotyledonen . . .* , plate 6 (fig. 9.1). S. Douady (fig. 9.2). Photo of Simon Schwendener, 1899, via Wikimedia Commons (fig. 9.3). Schwendener, plate 14 (fig. 9.4). Left, Schwendener, plate 10; center and right, S. Douady (fig. 9.5). Left, Schwendener, plate 10; right, S. Douady (fig. 9.6). Left, Schwendener, plate 8; right, S. Douady (fig. 9.7). Schwendener, plate 15 (fig. 9.8). S. Douady (fig. 9.9). Schwendener, plate 1 (fig. 9.10). Schwendener, plate 17 (fig. 9.11).

Gerrit van Iterson, *Mathematische und mikroskopisch-anatomische Studien über Blattstellungen*, plate 3, slightly modified for clarity (fig. 10.1). Photo of Van Iterson used by kind permission of Delft School of Microbiology Archives, Delft University of Technology, the Netherlands (fig. 10.2). Left, Van Iterson, plate 1; right, Van Iterson, 68 (fig. 10.3). Van Iterson, plate 3, detail (fig. 10.4). Van Iterson, plate 1, details, with graph by S. Douady and C. Golé (fig. 10.5). S. Douady (fig. 10.6). Van Iterson, 256 (fig. 10.7). Van Iterson, plate 5 (fig. 10.8). Van Iterson, plate 13, annotated by S. Douady and C. Golé (fig. 10.9). Van Iterson, 265 (fig. 10.10). Arthur

Harry Church illustrations, used by permission of The Natural History Museum/Alamy Stock Photos (figs. 10.11, 10.12).

Except where noted, all images in chapter 11 used by kind permission of the Provost and Scholars of King's College, Cambridge, UK. Photo of Alan Turing, King's College archives, University of Cambridge, AMT/K/7 12 (fig. 11.1). Drawing by Ethel Sara Turing, used by kind permission of Sherborne School Archives (fig. 11.2). Alan Turing to John Zachary Young, February 8, 1951, AMT/K/1 78a (fig. 11.3). AMT/K/3/11 (fig. 11.4). AMT/C/25 95 (fig. 11.5). AMT/K/3/5 (fig. 11.6). Left, University of Manchester Library, NAHC/TUR/C 2; right, University of Manchester Library, NAHC/TUR/C 3 (fig. 11.7). AMT/C/25 96 (fig. 11.8). S. Douady (fig. 11.9). Worksheet used by kind permission of Erinma Ochu and the Board of Trustees of the Science and Industry Museum, Manchester, UK (figs. 11.10, 11.11).

Arthur Veen and Aristid Lindenmayer, "Diffusion Mechanism for Phyllotaxis," *Plant Physiology* 60, no. 1, p. 134, recolored, used by permission of Oxford University Press (fig. 12.1). Veen and Lindenmayer, 133, used by permission of Oxford University Press (fig. 12.2). Aristid Lindenmayer and Przemysław Prusinkiewicz, *The Algorithmic Beauty of Plants*, 103, used by permission of Springer Nature (fig. 12.3). Hyun-Woo Lee and Leonid S. Levitov, "Universality in Phyllotaxis: A Mechanical Theory," used by permission of World Scientific Publishing (fig. 12.4). Lee and Levitov, 12, used by permission of World Scientific Publishing (fig. 12.5). Redux of fig. 10.1, slightly modified for clarity (fig. 12.6).

*Le Chercheur fantôme*, 54, used by permission of Robin Cousin and FLBLB (fig. 13.1). S. Douady (fig. 13.2). Philippe van Tieghem, *Traité de botanique*, 54 (fig. 13.3). S. Douady (figs. 13.4–13.11).

Poster used by kind permission of Matthieu Berthod and the Conservatoire et Jardin botaniques de la Ville de Genève, Switzerland (fig. 14.1). C. Golé (figs. 14.2, 14.3). SEM image by Jacques Dumais, annotated (fig. 14.4). C. Golé (fig. 14.5). SEM image used by kind permission of Rudolf Rutishauser, annotated by S. Douady (fig. 14.6). S. Douady (figs. 14.7–14.9). S. Douady and C. Golé (fig. 14.10). Left, drawing by A. Weisse in Karl von Goebel's *Organography of Plants*, 78; center, C. Golé; right, used by kind permission of Joanne Muni Douady Cornelissen (fig. 14.11). C. Golé (figs. 14.12–14.14).

S. Douady (figs. 15.1–15.3). C. Golé using *Mathematica* software (figs. 15.4, 15.5). S. Douady (figs. 15.6, 15.7). Redux of fig. 15.1 (fig. 15.8). S. Douady (fig. 15.9). Redux of fig. 10.1, slightly modified for clarity (fig. 15.10).

Used by kind permission of Etienne Couturier (fig. 16.1). Left, S. Douady; right, S. Douady, based on the work of E. Couturier (fig. 16.2). S. Douady (fig. 16.3). S. Douady, based on the work of E. Couturier (fig. 16.4). S. Douady (fig. 16.5). Used by kind permission of E. Couturier (fig. 16.6). Ambroise Tardieu, portrait of Michel Adanson, via Wikimedia Commons (fig. 16.7). Illustration in John Lubbock's *On Buds and Stipules*, 98 (fig. 16.8). S. Douady (figs. 16.9, 16.10).

Photo of Julius von Sachs, via X (formerly Twitter), (fig. 17.1). Left, filtered image by Jacques Dumais, based on Ralph Erickson's 1973 article "Tubular Packing of Spheres in Biological Fine Structure," *Science* 181, no. 4101, 713, fig. 6A, used by permission of the American Association for the Advancement of Science; right, Louis Bravais and Auguste Bravais, plate 2 (fig. 17.2). Erickson, fig. 4A, used by permission of the American Association for the Advancement of Science (fig. 17.3). Microscope image by Carlos S. Galvan-Ampudia et al., recolored, in "Temporal Integration of Auxin

Information for the Regulation of Patterning," via *eLife* (fig. 17.4). Redrawn by J. Dumais, based on J. C. Schoute, "Beiträge zur Blattstellungslehre," *Recueil des travaux botaniques néerlandais* 10, no. 3/4, figs. 1, 2, 9a (fig. 17.5).

Kajetan Zwieniecki, annotated by J. Dumais, used by permission (fig. 18.1). Photo of Léo Errera, *Recueil d'œuvre de Léo Errera*, via Biodiversity Heritage Library (fig. 18.2). Left, image redrawn by Jacques Dumais, based on D'Arcy Wentworth Thompson illustrations in *On Growth and Form*, figs. 153A and B; right, SEM image by J. Dumais (fig. 18.3). Left and right, bright-field microscope image and pictures based on J. Dumais, "Pattern Formation in Plants," *Current Opinion in Plant Biology* 10, p. 59, used by permission of Elsevier (fig. 18.4). J. Dumais (fig. 18.5). Left, photo by Stephen Petegorsky; right, bright-field microscope image by David W. Bierhorst, "On the Stem Apex, Leaf Initiation and Early Leaf Ontogeny in Filicalean Ferns," *American Journal of Botany* 64, no. 2, 126, recolored by J. Dumais, used by permission of John Wiley and Sons (fig. 18.6). Bright-field microscope image by Robert Geoffrion, used by permission (fig. 18.7).

Victor Mozqueda (fig. 19.1). S. Douady (fig. 19.2). Stewart Macdonald, "Arafura file snake (*Acrochordus arafurae*) in captivity," 2008, via Wikimedia Commons, CC BY-SA 3.0 (fig. 19.3). Sin Syue Li, "*Achalinus formosanus formosanus* full body shot," 2008, via Wikimedia Commons, CC BY 2.0 (fig. 19.4). Marlon Bunday, "Peacock in Full Display," 2014, via Wikimedia Commons, CC BY 3.0 (fig. 19.5). Fernando Losada Rodríguez, "Boca de lamprea.1 - Aquarium Finisterrae," 2007, via Wikimedia Commons, CC BY-SA 4.0 (fig. 19.6). SEM images by J. Dumais, annotated by J. Dumais and S. Douady (fig. 19.7).

S. Douady (fig. 20.1).

Victor Mozqueda (fig. 21.1). S. Douady (fig. 21.2). Victor Mozqueda (figs. 21.3–21.5).

Detail of fig. 2.1 (fig. A.1). Leonardo da Vinci, Notebook M, 78v, used by permission of Alamy Stock Photos (fig. A.2). Left, image from Johannes Kepler, *The Six-Cornered Snowflake*; right, S. Douady (fig. A.3). Left, detail of Kepler image shown at right; right, Kepler, "Kepler Platonic solids," from *Mysterium Cosmographicum*, via Wikimedia Commons (fig. A.4). Bonnet, plate 21 (fig. A.5). Bonnet, plates 22 and 23 (fig. A.6). Aaron Rotenberg, "SternBrocotTree," 2008, via Wikimedia Commons, CC BY-SA 3.0 (fig. A.7). S. Douady (fig. A.8). Braun, plate 11 (fig. A.9). Braun, plate 10 (fig. A.10). Braun, plate 22, annotated by S. Douady (fig. A.11). Braun, plate 50, detail, annotated by S. Douady (fig. A.12). Redux of fig. 12.4 (fig. A.13). Redux of fig. 12.5 (fig. A.14). Redux of fig. 13.7 (fig. A.15). Redux of fig. 13.8 (fig. A.16). Redux of fig. 13.9 (fig. A.17). Redux of fig. 13.11 (fig. A.18). Image by S. Douady (fig. A.19). C. Golé (fig. A.20). S. Douady and C. Golé (figs. A.21, A22). C. Golé (fig. A.23). P. Atela, C. Golé, and S. Hotton, "A Dynamical System for Plant Pattern Formation," *Journal of Nonlinear Science* 12, no. 6, p. 645, used by permission of Springer Nature (fig. A.24). "Hyperbolic domains 642," via Wikimedia Commons (fig. A.25).

# ACKNOWLEDGMENTS

**SD:** I would like to thank my great-grandfather Rémi Perrier, for buying a botanical book, Yves Couder for allowing me to work on plants during my military service in his lab, Etienne Couturier for his genuine observations, Christophe Golé for his stimulating friendship, Jacques Dumais for his critical remarks, Annemiek Cornelissen, Joanne and Anand, our kids, for supporting me through the delivery of this book to the world, and Nancy Pick for her unwavering will to make this book readable, beautiful, and finished. Thanks to the world for its marvelous plants and for giving this body the ability to perceive them.

**JD:** For a young scientist in search of order in Nature, phyllotaxis provided a secluded playground, one accessible only to those willing to pause their hurried lives and devote their attention to minute details of plants. As luck would have it, on this botanical playground I forged many of my most significant research collaborations, starting with my coauthors Christophe Golé and Stéphane Douady, whose friendship and perspectives on phyllotaxis have left an indelible mark on me. Yet I doubt this book would have become a reality without Nancy Pick at the helm. Her patient hand and positive outlook guided our work for more than a decade.

I would also like to thank the many students and young scientists who have worked with me on questions relating to phyllotaxis and cell division. Their dedication and enthusiasm have made the journey truly enjoyable. My most profound gratitude is reserved for my wife, Mariela Cerda, who, for more than two decades, has had to make do with a husband whose mind would too often wander off into a distant botanical playground.

**CG:** I would like to thank Scott Hotton for introducing me to the limitless world of phyllotaxis. My gratitude to Pau Atela, Michael Marcotrigiano and Madeleine Zadik, who conspired with me to create the Plant Spirals exhibit at Smith College. Through this endeavor, I had the good fortune to meet my coauthors and friends Stéphane Douady and Jacques Dumais. Denise Lello has been a tremendous help in my learning the little botany I know and in shepherding the creation of the Plant-Math lab at Smith College. I can't begin to thank all the many students who have participated in research with us over the years. I am grateful to each one of them for the inspiration they continue to give me. Thanks to Rob Dorit, Christophe Godin, Peter Schlessinger, Marguerite Suozzo-Golé and Liz Suozzo for their collaboration, support and many inspiring discussions. And of course, I still can't believe my luck in meeting Nancy Pick some 14 years ago, and that we would be crazy enough to decide within minutes to start this book project. She's the magical fairy who has given a voice to this book.

**NP:** This book was a wild, dynamical collaboration that took a Fibonacci number of years to complete: 13. We coauthors wish to thank Vickie Kearn, our first editor, for taking a chance on us. Her successor, Diana Gillooly, brought clarity and fierce intelligence to this project, and we are deeply grateful for her guidance. Inspired by Fibonacci ratios, book designer Chris Ferrante made these pages beautiful, as did the gorgeous botanical photos of Victor Mozqueda. Through the years, I was sustained by Amy Gray's insights and by many dear friends, especially Alison Sparks and Ilan Stavans, Betsy Kolbert and John Kleiner, Susan Faludi and Russ Rymer, Juliette Bogaers, Lo Dagerman, and Marta Siberio. My hubby of nearly 34 (!) years, Lawrence Douglas, fed me spaghetti and love, and our amazing sons, Jacob and Milo Douglas, made me laugh at my spiraling obsessions. Lastly, I am grateful to my endlessly inventive coauthors Christophe, Jacques, and Stéphane for their kindness through thick and thin, and for opening my eyes.

# INDEX

acorn caps, 203, 218
Adanson, Michel, 239–40; portrait of, 240
Adler, Irving, 196–98
aloe plant, 5, 6
alternate phyllotaxis, 58, 72
alternation of generations, 119–21
animals, 265–73
*Anthurium*, 135, 209, 213, 279
apical meristem: microscope images of, 11, 121–26, 205–6, 208; mother cell of, 261, 263; Turing's analysis of, 168, 169. *See also* meristem
Araceae family, 135–38, 218, 278–79
Archimedean spirals, 32, 86, 315n8(Ch.6)
artichokes, 205–7, 283
asparagus, 284
Atela, Pau, 204, 320n1
attractors, 261, 263
auxin, 251–53

*Banksia*, 217
Besson, Sébastien, 259
bifurcations: broken, 194; of cylindrical protein lattices, 250; Van Iterson diagram and, 181–82, 194–96, 320n5
bijugate phyllotaxis, 137, 303
biomechanics of plants, 131–32

Bonnet, Charles, 53–66; on dew, 54–57, 58, 312n6, 312n8; exceptions to leaf orders of, 63–64, 292–94, 313n22; five leaf orders of, 58–66; portrait of, 54; Schimper inspired by, 69, 70–72
Bonnet syndrome, 64
Braun, Alexander, 68–69, 72, 76–77, 78–90; discoveries of, 112–13. *See also* Schimper and Braun's spiral theory of phyllotaxis
Braun, Cecilie, 79–81, 88; portrait of, 79
Bravais brothers: discoveries of, 112–13; fieldwork of, 106–8; golden angle and, 88, 96–97, 100–103, 106–10, 112, 149; illustrations drawn by, 100–101, 103–4; lattices of, 103–6, 180, 182, 315n7; lives of, 96–99; portrait of Auguste, 97
Bravais lattices, in crystallography, 97, 110, 112, 278
buds: apical meristem of, 122–24, 168, 169; cabbage as, 234–35; packing of leaves inside, 127–28, 233–43, 273, 323n2; of radish, 86, 88; Turing's analysis of growth in, 168, 169
bur chervil, 226, 227, 231

cabbage, 215–16, 234–36, 282
Calandrini, Jean-Louis, 54, 58, 59, 62, 64, 312n16

camera lucida, 137
Cantor, Georg, 220–21
Cantor set, 221, 224
cell division, 254–64
cell walls, 256
chemical morphogenesis, 164–66, 168, 169
Church, Arthur Harry, 145, 156–57
climate, and leaf shape, 239
computer simulations: disk stacking in, 129–30, 134, 136, 205; finding parameters in, 192–99; Lindenmayer's early results, 176–79; Snow and Snow hypothesis in, 303; Turing's first for phyllotaxis, 165
contact pressure, 139–41, 182, 187, 252–53
continued fractions, 68, 105–6, 108–10
corn, 12, 64, 200, 203, 216, 292, 321n6
Couder, Yves, 170, 183, 186–87, 189, 191–92, 198, 204
Couturier, Etienne, 233–34, 238–39
cylindrical (cylinder) lattice, 10–11, 112, 249–50

dahlia, 13, 92–95
Darwin, Charles, 272–73
decussate phyllotaxis, 38, 58
deformation, 132, 139–41, 182. *See also* contact pressure
disk diameter: quasi-symmetric patterns and, 216–17; Van Iterson diagram and, 147, 153–54
disk lattices. *See* lattices
disk stacking, 129–30, 133–36, 149, 155, 176, 205, 316n12. *See also* rhombic tilings
divergence angle: of Braun's pinecones, 82–83, 84–86, 314n6; in cell division model, 262; coined by Schimper, 69, 112; constant and irrational, 102–6; defined, 8; generative spirals and, 9–10, 104–6; as quotient of Fibonacci numbers, 72–76; rational for Schimper and Braun, 102–3, 149; in Schimper's watch face model, 70–72; Van Iterson diagram and, 147, 149, 262. *See also* golden angle
divine proportion, 45, 49. *See also* golden ratio φ
dodecahedron, 45, 48
Douady rabbit, 224–25, 307
dynamical systems, 12, 126, 166, 187–88, 195–97, 261

empiricists vs. idealists, 247–48, 250–51
energy in Levitov's analysis, 179–81, 299–300, 320n10
Erickson, Ralph O., 250
Errera, Léo, 255–56, 257–58, 259, 260, 264; portrait of, 256

Farey sequence, 278, 294, 295
ferns: alternation of generations, 119–20; apical cell of, 263; Archimedean spiral and, 32, 33
Fibonacci (Leonardo de Pisa), 24–28
Fibonacci numbers: in Bonnet's observations, 63–64; defined, 7; divergence angles as quotients of, 72–76; sunflowers not conforming to, 172; in Van Iterson's diagram, 149–50. *See also* parastichy numbers
Fibonacci poem, 77, 309n1(Ch.1)
Fibonacci sequence, 1; of bee drone ancestors, 31; defined, 7; golden ratio and, 45–46, 48, 311nn5–6; of Indian poets, 28–30; Kepler and, 41, 45–46, 48; Leonardo de Pisa and, 24–28; in rabbit problem, 26–28
Fibonacci spirals: in animals, 266, 271; functional significance of, 273, 325n16; hands-on activities, 31–33, 65–67; in magnetic droplets, 190–91; transitions between, 155
fir trees, 39–40, 141
flowers: five-petaled, 41, 43–44, 46; labeling floret numbers, 92–94

formation, and Schwendener's biomechanics, 132, 139, 182
fractals, 219–24; in plants, 224, 226, 231–32. *See also* self-similarity
fraction trees, 74–75, 84–86, 294–96
fundamental hypothesis of phyllotaxis, 180–81, 319n10
Fundamental Theorem of Phyllotaxis, 103

generative spirals: of Bravais brothers, 104–6; defined, 9–10; left behind by Hofmeister's rule, 133; Schimper's construction of, 72, 112
Goethe, Johann Wolfgang von, 120, 122
golden angle, 8, 47; Bravais brothers and, 88, 96–97, 102–3, 106–10, 112, 149; continued fractions and, 108–10; as divergence angle, 8, 100, 102–3, 108, 149; hands-on activity, 113–15
golden mean. *See* golden ratio φ
golden ratio φ, 45–49; Bravais brothers and, 110; Kepler and, 41, 42, 46; myth about, 48–49; renormalization and, 155
golden section. *See* golden ratio φ
grasses, 38
Greece, Ancient, 22
growth parameter, 192–94

Hakim, Vincent, 194
Han Ying, 22
helix (helices), 58, 62, 66, 72, 81, 309n4(Intro.)
Hemachandra, 28–29
hexagonal lattices: in evolution, 273; Levitov and, 180, 299–300; in transitions, 306; in Van Iterson tree, 149–51, 318n10, 320n8
hexagonal packings, 149, 180, 273, 299. *See also* hexagonal lattices
hexagonal patterns in animals, 265, 267–70, 324nn5–7(Ch.19)
Hofmeister, Wilhelm, 119–28, 277; portrait of, 121

Hofmeister's rule, 122, 124–26; disk stacking and, 133–34; empiricists' use of, 248; magnetic droplets and, 188, 198; transitions and, 212
honeybee drone genealogy, 31
horsetails, 12
Hotton, Scott, 204, 305, 320n1
hyperbolic geometry, 308, 317n9, 326n17

icosahedron, 45
idealists vs. empiricists, 247–48, 250–51
Indian poets, 28–30
inhibitor, and pattern formation, 164–65, 176–77, 251–53
irregular cases, 278. *See also* quasi-symmetric phyllotaxis

Kepler, Johannes, 22, 41–44, 290–91; portrait of, 42
*kirigami*, 233, 235, 238, 241–43
Koch snowflake, 222–23, 224

lattices, 10–11; of Bravais brothers' phyllotaxis, 103–6, 180, 182, 315n7; of Bravais crystallography, 110, 112; in Schwendener's deformation model, 139–41; of Van Iterson, 146–50, 153–55
leaves folded in buds, 127–28, 233–43, 323n2(Ch.16)
Leonardo da Vinci, 34–40, 45, 49, 311nn5–7; quincunx and, 38, 58; trees and, 39–40, 289–90, 325n2
Leonardo de Pisa. *See* Fibonacci
Levitov, Leonid S., 179–82, 185, 187, 194, 195–98; energy and, 179–81, 299–300, 320n10
lichens, 130–31
Lindenmayer, Aristid, 171, 176–79, 226, 228
logarithmic spirals, 32, 33, 271
L-systems, 127, 156, 176, 178, 228, 319n2
Lubbock, John, 240
Lucas, Édouard, 28

magnetic droplets, 182, 187–91, 198
magnolias, 135–36, 137
Mandelbrot, Benoit B., 223–24
Mandelbrot set, 204, 224, 225, 304–05
mediant, 75, 79, 86, 108–9, 294–96
meristem, 11; Hofmeister's rule and, 248; optimal packing in, 273; zigzag fronts and, 208–9. *See also* apical meristem
microscope: apical meristem and, 11, 121–26, 205–6, 208; cell division and, 257; Hofmeister and, 120, 121–24; Schwendener and, 129, 130–31, 133, 137
microtubules, 263–64
mollusks, spirals in, 32–33, 270–71
morphogenesis: chemical, 164–66, 168, 169; growth history and, 198; patterns emerging from, 277
myrtle, 22, 23, 60

Nägeli, Carl von, 121–22
Nagpal, Radhika, 261
natural selection, 272–73
nautilus shell, 33

orthostichies, 84–85, 86, 112, 326n11. *See also* vertical spirals

parastichy: Braun's introduction of, 81; defined, 7
parastichy numbers, 7; as adjacent Fibonacci numbers, 108; Braun's introduction of, 83; in computer simulations, 179, 192–93; counting for yourself, 13–17; double Fibonacci numbers in, 137; with magnetic droplets, 189–91; nearly equal, 216–18; relatively prime, 95, 105; starting at (1,1) or (2,2), 214–15; Turing's observation of, 166–67; Van Iterson diagram and, 146–50; zigzag fronts and, 209–10. *See also* transitions in parastichy numbers
peace lily, 135, 210–11, 212, 218, 279

pentagons, 45–48; in Bonnet's quincunx order, 62; in ommatidia pattern, 270; transitions and, 203, 211–12
*petai*, 217
photosynthesis, 53, 57
phyllotaxis: coined by Schimper, 69; duality of Fibonacci or quasi-symmetric, 218; era of fixed geometry, 112–13; Greek for leaf arrangement, 2, 69
pineapples, 13–14, 281
pinecones: Braun's work on, 76–77, 78–87, 112, 296–99, 314n6; counting spirals in, 4–5, 7, 15–17, 111, 186; divergence angles of, 82–83, 84–86, 314n6(Ch.6); numbering scales of, 80, 81, 83; radial view of, 86–87; vertical spirals in, 82–83, 84, 112
Plantefol, Lucien, 187
Plateau, Joseph, 255–56
Pliny the Elder, 22, 24
primordia, 11, 122–26; added singly by simple rules, 176; bifurcations and, 194–96; formed in L-systems, 176–77; Hofmeister's rule and, 122, 124–26; Schwendener's disk stacking and, 133–35; Turing's analysis of, 164–66, 168–69; Van Iterson diagram and, 147, 149, 153, 155; zigzag fronts and, 208–9
protein crystals, 248–51
Prusinkiewicz, Przemyslaw, 178–79, 228

quasi-symmetric phyllotaxis, 12, 203, 216–18, 273, 279
quincunx: Bonnet's description of, 53, 58, 61–62; Bonnet's variations of, 63–64, 292–93, 313n22; cardboard model of, 65; divergence angle of, 72, 74; Leonardo's observation of, 38

rabbit problem, 26–28
recipes, 281–87
redoubled spirals, 53, 58, 62–64, 66, 72
renormalization, 153–55, 278, 308

rhombic tilings, 205, 321n3
rhombuses in transitions, 203, 211–12
rise, 193, 262
Romanesco broccoli, 183–86; fractal view of, 220, 226, 228–29, 231; recipe for, 285

Sanskrit poets, 28–30
Schimper, Karl, 68–77, 78–79, 86; discoveries of, 112–13; portrait of, 69
Schimper and Braun's spiral theory of phyllotaxis, 79, 88; Bravais brothers and, 98, 102–3; dismissed by Hofmeister, 121
Schoute, Johannes Cornelis, 252
Schwendener, Simon, 122, 126, 129–41; contact pressure theory of, 139–41, 182, 187, 252–53; disk stacking and, 133–34; microscope and, 129, 130–31, 133, 137; portrait of, 131; unusual transitions and, 135–36
self-similarity: in Church's art, 156; of Van Iterson's diagram, 147, 149, 153, 182, 220, 230–31, 307–8. *See also* fractals
shoot apical meristem (SAM), 11. *See also* apical meristem
Sierpiński carpet, 221–22
Snow and Snow rule, 198, 303
soap bubbles, 254–64
spiral lattice, 10–11, 299
spirals in plants, 5–6; Bonnet and Calandrini as first to write about, 53, 62–64; in cabbage leaves, 234–35, 236; Fibonacci spirals, 31–33, 65–67; Leonardo's observation of, 38, 40. *See also* parastichy numbers; Schimper and Braun's spiral theory of phyllotaxis
spruce branch, 208–9
square lattices, 92, 149–51, 154, 196, 270, 273, 299–300, 306, 318n10, 320n10, 324n5(Ch.19)
strawberries, 12, 135, 203, 210, 218, 286
sunflowers: computer-generated, 178; counting spirals in, 4–5, 7, 172–74; dynamic construction of, 198; parastichy numbers of, 108, 215; Turing's work on, 166–67, 172
sunlight on leaves, 5–6, 273
Swinton, Jonathan, 167–68, 170, 172
sycamore maple, 235, 236, 238, 240

tangrams, and cell division, 254, 263
teasels, 135, 137–38
Theophrastus, 22, 23
Thompson, D'Arcy Wentworth, 143–44, 256–58, 259, 260
threshold of dynamical stability, 195–97
transitions in parastichy numbers, 130, 134–36; in pineapples, 13; Turing's analysis of, 169–70; Van Iterson's analysis and, 153–55; zigzag fronts and, 201, 203, 206–7, 210–16
transitions in quasi-symmetric patterns, 216–18
trees: leaves folded in buds, 235–40; Leonardo on, 39–40, 289–90, 325n2
triangle transitions, 203, 210–17, 305
Try Your Hand: Fibonacci poem, 77; Fibonacci sequences, 31–33; Fibonacci stem, 91–92; golden angle of divergence, 113–15; *kirigami* maple leaf, 241–43; labeling floret numbers, 92–94; parastichies, 13–17; spiral stems, 65–67; Turing's sunflowers, 172–73
Turing, Alan, 95, 161–74; chemical patterns and, 164–66, 251; death of, 171; dynamical systems and, 126, 166; hypothesis of geometric phyllotaxis, 170, 180–81, 182; portrait of, 162
Turing instability, 166

Van Iterson, Gerrit, 130, 132, 142–56; portrait of, 144
Van Iterson diagram, 142–43, 146–50; cell division and, 254, 261–62; computer simulations and, 178, 179–81, 193–97; Fibonacci branch of, 149, 150, 155, 179; self-similarity of, 147, 149, 153, 182, 220, 230–31, 307–8

Veen, Arthur, 176–78
Venus flytrap, 257–60
vertical spirals: in corn, 200; oscillating a little, 314n5; in pinecones, 82–83, 84, 112. *See also* orthostichies
von Sachs, Julius, 247–48; portrait of, 248

whorled phyllotaxis, 12; Bonnet's observations of, 58; ignored in simple model, 204–5; Leonardo's observations of, 38, 39–40; noticed by the ancients, 22–24; in Schwendener's magnolia drawing, 136, 317n13
Wiener, Norbert, 258–59, 263

yellow asphodel, 106–8
Young, John Zachary, 163–64

zigzag fronts: counting parastichies with, 209–10; emergence of, 208–9; transitions and, 201, 203, 206–7, 210–16
zigzag line of Van Iterson, 155, 321n4